A SCIENCE ODYSSEY

A SCIENCE ODYSSEY

ODYSSEY

100 YEARS OF DISCOVERY

Charles Flowers

Foreword by Charles Osgood

Introduction by Charles Kuralt

WILLIAM MORROW AND COMPANY, INC.

New York

Overleaf: Several hundred galaxies never before seen, perhaps four billion times fainter than the eye can detect, may lie at the very edge of the visible horizon of the universe in this "deepest ever" space photo.

Library of Congress Cataloging-in-Publication Data

Flowers, Charles.
    A science odyssey : 100 years of discovery / Charles Flowers ;
  introduction by Charles Kuralt.
        p.   cm.
    Includes index.
    ISBN 0-688-15196-5
    1. Science—History—20th century.   2. Discoveries in science—History—
  20th century.   3. Technology—History—20th century.
  I. Title.
  Q125.F592   1998
  509'.04—dc21   97-35745
                                                                                CIP

Printed in the United States of America

First Edition

1 2 3 4 5 6 7 8 9 10

BOOK DESIGN BY GAYE KORBET, WGBH DESIGN, AND NATALIE PANGARO

www.williammorrow.com

To Howard Fischer Flowers, 1920-1997

ACKNOWLEDGMENTS

WGBH is grateful for the efforts of so many who helped make this book possible. Just as the book tells what television cannot show, the programs reveal the century in ways a book never can. To this end, it has been our goal to create book and television as true companions to each other.

Few people other than an author and editor can ever fully appreciate what goes into a collaborative book of this type. It requires mastery of sometimes contradictory and almost always grueling tasks of creativity, translation, research, and coordination. Author Charles Flowers created this lovely and far-ranging account of the twentieth century in part out of his own resources and in part out of the stories, ideas, and images gathered for the WGBH-produced PBS television series *A Science Odyssey* and its related national educational outreach initiative. His wit and style are everywhere evident here, as were his fortitude and good humor during the project, for all of which we owe him a deep debt of thanks. Editor Nancy Lattanzio, too, has our respect and gratitude for keeping the book on track in all respects. Theirs has been a daunting chore, beautifully realized.

Thanks are also due to WGBH photo researcher Susan Levene, WGBH designer Gaye Korbet, designer Natalie Pangaro, and WGBH production assistant Robert Banks, for their hard work and professionalism, and, especially, to Senior Editor Paul Bresnick of William Morrow for constant support.

We owe a special debt to the producers of the television programs, the wellspring for the book. Developed over a period of five years, the series was specifically designed to expand public understanding of the nature of science in our own time. The book has benefitted greatly from this effort and from the incisive and finely crafted television that has come out of it. Thank you, first and foremost, to Executive Producer Thomas Friedman, and Executive in Charge and NOVA Executive Producer Paula Apsell.

Individual program producers and other key series staff included Noel Buckner and Rob Whittlesey, producers, and Elizabeth Carver, associate producer, "Physics and

Astronomy"; Carl Charlson, producer, and Harry Gural, associate producer, "Technology"; David Espar and Michael Rossiter, producers, Diane Hendrix, coproducer, and Angela Spindler-Brown, associate producer, "Origins"; Larry Klein, producer, Sarah Holt, coproducer, and Catherine Sager, associate producer, "Medicine"; Alice Markowitz, producer, and Cheryl Gall, coproducer, "Human Behavior"; Lisa Mirowitz, series coordinating producer; Lisa D'Angelo, series associate producer; Kate Halpern, series production assistant; Julie Benyo, outreach director; and Kathleen Shugrue, director of administration.

The expertise and scrutiny of several outside readers was of substantial help to the book: David Kaiser, Harvard University, History of Science Department, and James Ray, Annexe Montpellier, for Chapter One; Steven Lubar, the American Museum of Natural History, for Chapter Two; Greg Laden, Harvard University, Department of Anthropology, and James Cullen, Salem State University, Department of Geology, for Chapter Three; Colin Talley, University of California at San Francisco, for Chapter Four; Nadine Weidman, Harvard University, History of Science Department, and Eric Hollander, College of Physicians and Surgeons of Columbia University, Department of Psychiatry, for Chapter Five.

Without the commitment and vision of the funders, the television series and outreach would not have been possible. Grateful thanks to them:

Major funding was provided by the National Science Foundation.

Corporate sponsorship was provided by IBM.

Additional funding was provided by public television viewers, the Corporation for Public Broadcasting, the Arthur Vining Davis Foundations, the Carnegie Corporation of New York, and Becton Dickinson and Company.

Charles Osgood, series host, rose heroically to the challenge of writing a perfect foreword to the book on extremely short notice.

Finally, we are grateful to the late Charles Kuralt, who worked with us on introductory material for the book as long as he was physically able.

— KAREN JOHNSON
Executive Director, Publishing and Products
WGBH

Using electrical charges to separate DNA molecules by size, as in these bands shown in a digitally colored photo, is the first step in typing a cancer or identifying a child's father.

CONTENTS

*Electric lights, the telephone, the radio—
they lightened life at home, brought the
world to the table, then unpredictably
unleashed the flood of knowledge and
technology that has made our century
unlike any other.*

FOREWORD

The more we come to know, the more we realize how little we know. The more we understand, the more clear it is that everything we have learned is nothing compared to what we have yet to learn. Behind each locked door we have managed to open are still more doors and more locks, and so on *ad infinitum.* So science is not an arrival, but a journey. It is not a fixed body of knowledge or growing shelf of facts and theories, but an infinite series of questions. The most brilliant scientists have been those who have sought not the right answers to give but the right questions to ask.

And so we have progressed, not in a straight line, but from ignorance to misconception, and from misconception to mistake, and from mistake to failure, and from failure to insight, and from insight to discovery about ourselves and our universe and how things work. It happened very slowly and haltingly at first and then faster and faster, picking up tempo until in the twentieth century it has reached a dizzying speed. In my own lifetime there have been revolutions in just about every branch of human knowledge. I wonder how many pages in my high school science books still stand. Today we laugh at how little we knew yesterday. Tomorrow we will laugh at how little we know today.

There are ordinary high school kids who can tell you now about strange stars and invisible forces in the universe that even the brightest astronomer knew nothing about only a few decades ago. Kids today know about an expanding universe and the explosion that started it, the blast furnace of creation, and *when* and *how* it most likely happened. They know, as our grandfathers did not, that the continents are constantly on the move, fixed to vast plates that slide over and bang against one another, rocking the earth. Kids today can tell you that more than 90 percent of the universe is so-called dark matter, invisible and unknowable. When I was a boy, radio still seemed a technological wonder and space exploration was something in the "Buck Rogers" comic strip. Today's children take television, computers, and cellular telephones for granted, as they do space stations and missions to Mars.

Science and technology have touched all of our lives and changed us in ways that we ourselves cannot yet comprehend. Lifesaving medical miracles have become so commonplace that we are impatient about them. We don't want to wait. We want cures for AIDS and cancer and other diseases, and we want them now! Perhaps we have come to expect too much from science. We look to it for the answers when it is still busy with the questions. Yet we distrust and fear science and technology too, and it is not unreasonable that we should. For all its benefits, we have seen its destructive side as well. We understand that it can be a wonderful servant but a terrible master.

All of which is by way of an introduction to an introduction to *A Science Odyssey,* this book and the PBS television series on which it is based. It is about a century of physics and astronomy, and what we've learned in the last hundred years about the origins of the earth and life, medicine, and the exploration of the human mind and personality. My good friend and longtime CBS News colleague Charles Kuralt was to have been the host and guide for these broadcasts. Sad to say, Charles died on the Fourth of July, 1997, and there was nothing medical science could do to save him. We are going ahead with the trip even so, and I will do my best to show you what Charles wanted to show you, and tell you the stories he wanted you to hear. First, though, his own introduction in his own words.

—CHARLES OSGOOD
August 1997

INTRODUCTION

It was the only time I ever saw my grandmother cry. She was standing by the dirt road that passed our farm, wiping her eyes with a corner of her apron. This was alarming to me as a little boy. I went running barefoot to see what was the matter. It turned out that those tears of hers were tears of joy. She could see the Rural Electrification Administration light poles coming down the road toward our house. We were going to have electricity! We were going to enter the modern age!

There was no disgrace in those Depression years in living on a farm without electric power. In the year of my birth, 1934, electricity had reached only about one American farm in ten. Nobody we knew had electric lights, except for Uncle John, clerk of the court at the county seat, who lived in town. Nobody had indoor plumbing either. If you wanted a drink of water, you got it from a drinking gourd that hung beside the hand pump on the porch. The "bathroom" was in the backyard, a pungent outhouse, or, at night, a "slop jar" under the bed. Bathing was accomplished in a galvanized-iron tub set out beside the kitchen door. On laundry day, my grandmother used a hoe handle to stir a steaming mess of underwear and overalls and lye soap in a big black pot over an outdoor fire. When electricity finally arrived, our first use of it was to light a bare electric bulb my father rigged to hang over the kitchen table. The second was to power a water pump, so that my grandmother could do the laundry indoors in the sink.

We were not unaware of the marvels of the modern world that were changing people's lives in the cities. I remember seeing a color photograph of Times Square, a calendar illustration, that seemed impossibly glamorous, all tall buildings and taxicabs and hurrying people in hats and coats and ties—not a farm animal or an unpaved plot of ground anywhere to be seen. A relative actually took the train to the 1939 World's Fair in New York and came back with fabulous stories to tell, and a bronze Trylon and Perisphere paperweight as a gift for me. We knew about the odyssey of science; we just weren't part of it.

I am trying to think of examples of modern technology that I knew firsthand as a child in the thirties.

There was a magneto-operated party-line telephone hanging on the wall; it rang from time to time, but hardly ever with "our" ring, two shorts and a long. In fact, I can't remember anybody in the house talking on that phone. My grandfather hated the thing. If there was some neighboring farmer he wanted to talk to, he did what he had always done: open the screen door, walk out of the house and down the road, and see the man face-to-face.

There was a telephone. What else?

There was a wind-up Victrola with one-sided, scratchy recordings of Nelson Eddy, Harry Lauder, and Enrico Caruso, and a shiny round box built in beside the turntable to hold the steel needles. We changed the needles frequently; one worked as well as another, none of them very well.

And there was a radio. It was a battery-powered Montgomery Ward floor model in the shape of a Gothic arch, with a great round dial that glowed orange when the radio was turned on. Printed on the dial were not numbers but letters—the call letters of the radio stations of the day, KDKA, WGN, WCKY, and the others. Understand, none of these stations actually came in on the radio, but it was magical to think that some night they might. The only station we could be pretty sure of receiving was WPTF in Raleigh, the capital of our state. I don't remember anything that was said on the radio. I remember only that voices came through the air. I didn't know how it worked. (I *still* don't know how it works.)

These were all luxuries, of course. Everything that made that farm work was based on ancient technology indeed. A spinning wheel and a loom filled a side room. There, on winter afternoons, my mother and grandmother had long, affectionate talks while weaving shawls or bedspreads, just as women had done for millennia.

Before the great day when electricity arrived on the farm, I learned to read by the light of a kerosene lamp with a glass chimney, very much (except for the fuel) like the lamp by which children of colonial America learned to read—and only a slight improvement over the design of the lamp by which children learned to read in imperial Rome.

Cooking in that house, a more or less constant activity from morning to night, was done on a woodstove in the kitchen. What was cooked came from the garden—sweet potatoes, tomatoes, beets, butter beans; or from the fields—blackeyed peas, corn, and collard greens; or from the pantry—this same food in bright rows of Mason jars, put up for the winter along with apple butter and apple jelly. (No real farm of that day or this was without apple trees.)

Milk was provided by two milk cows, eggs by the chickens in the chicken coop (and boiled chicken on Sunday by some luckless hen that had stopped laying). And marvelous, hard, salt-cured country ham came from my grandfather's smokehouse. (I resisted thinking about where the ham *really* came from until, one autumn, the brutality and squealing and blood of hog-slaughtering day intruded on my childhood idyll. That was the day I resolved never to be a farmer.)

We lived pretty much outside the money economy. I know we bought staples like cornmeal, which came in flowered fifty-pound sacks intended to be turned into girls' blouses and skirts. And my grandparents must have also bought salt and sugar and the like. But nearly all our food was homegrown, and since we didn't raise beef cattle, I was considerably older before I tasted my first hamburger.

This self-sufficiency extended to clothing, although my grandmother's Sunday dress and her hats and shoes and my grandfather's boots must have been purchased in town. Nearly everything else was stitched up on a treadle-powered sewing machine (patented by Elias Howe about a hundred years earlier), probably the most useful thing in the house.

There were cast-iron wood-burning stoves for heating the bedrooms, but they were never used because they were too much trouble to keep lighted. Everybody slept under piles of quilts on cold nights, then dashed downstairs in the morning and dressed in the warm kitchen, which was the center of life, winter and summer.

The technology of the working farm, like most of that of the farmhouse, was centuries old—the technology of ax, plow, and harrow. My grandfather cultivated his fields by plodding behind a mule from dawn until dusk, as yeoman farmers always had done. He harvested his crops and brought them into the barn in a mule cart, then took them to market in a mule-drawn wagon. Later, an automobile came into the family— that of my aunt Trixie, who had come home ill from teachers college—but there was little use for a car in the settled life of the farm; you cannot use a Chevrolet to plow a furrow.

All my memories of that farm are happy ones. I was a child. At night, in the dark, I wondered at the universe of brilliant stars overhead; only a few times since, in other places devoid of artificial light, have I seen the stars so vividly. By day, I had a corncrib for rolling around in, a woodpile for making imaginary forts against Indian attacks, a hayloft for hiding out, and a sycamore tree for climbing. There were dangers to watch out for, too—sandspurs in the grass and spiders in the barn. There was even a snake in the apple orchard, which made the place, almost literally, a kind of Eden.

I would willingly go back to that place and time if I could be a child again. But I realize I would not wish it on any adult. For my grandparents, it was a place of unending and back-breaking physical labor, relieved only by my grandmother's mental exertions as a schoolteacher. Leisure was a concept alien to them both, and they could have had no hope for an easier future. I see now that they must have spent their lives in a state of near exhaustion, turning to despair as illness struck their family, and death. My mother was one of three sisters. The other two died in their twenties of a form of tuberculosis they would not contract today, and that, if they did, would easily be cured.

The odyssey of science arrived late on that farm, too late to provide my grandmother with so much as a dishwasher or vacuum cleaner, too late to erase my grandfather's burden with a chainsaw or a tractor. After my grandfather's death (also of tuberculosis), my grandmother left the farm to move in with us in a city far across the state. She lived long enough to enjoy reading for pleasure, as she had last been able to do as a child growing up in the late nineteenth century. Before she died, she came to take for granted the assorted miracles of the twentieth century: She traveled on a jet plane and discovered a couple of favorite television programs.

Once, we went back to see if we could find the farm. We could not. The dirt road that passed the house had become a paved four-lane boulevard. Among the fast-food restaurants and strip malls, we could not guess where my grandfather had once plowed his fields, or the old sycamores had once shaded the old house.

I was overcome by nostalgia for the scene of my childhood and dismay at what had become of it. I don't think grandmother felt that way.

"Well," she said, "that's progress for you."

My grandmother did not have an ironic turn of mind. She and I looked at the suburban jumble and saw different things. I saw asphalt and traffic and garish advertising signs. She saw ease and convenience. I think she compared modern life along that road to the life she had lived there, and meant what she said:

"That's progress."

—Charles Kuralt

June 1997

A SCIENCE ODYSSEY

 CHAPTER ONE

MYSTERIES OF THE UNIVERSE

An odyssey is not a planned trip from one place to another. The ship sets sail, but its course is determined by the chance sighting of an exotic isle, the sudden crash of a hurricane, the mistakes or foibles of the pilot at the helm.

In the beginning of the twentieth century, the explorations of scientists and philosophers, physicians and thinkers, which had only slowly progressed throughout the ages, suddenly accelerated with exhilarating speed. Without anyone's fully recognizing what would happen, the world of our ancestors began changing in unpredictable ways—some exciting, some terrifying, but none reversible.

Looking to clarify the stars shining on earth, we found ourselves a speck in a vast, unimaginably expanding universe; hoping to cure mental disease, we stumbled upon a thousand troubling questions about the nature of personality and identity and free will; expecting to create better lives for all, we have sometimes created unprecedented ills.

This odyssey goes on, because we cannot resist the thrill of learning what could never before be imagined. Knowledge for its own sake is food enough, not just for scientists. We all share in the delights of surprising discoveries, even when they shatter our most ancient, cherished ideas about ourselves and the nature of the universe. The mind reels, but the blood races. What next?

The odyssey of science is irresistible.

Called "Little Cloud" by the Persian astronomer al-Sufi, who discovered it in A.D. 950, the Andromeda nebula with its more than three hundred billion stars gravitates around our own Milky Way at the relatively close distance of 2,200,000 light-years. Until early in the twentieth century, it was thought to be a gaseous glow within our galaxy, where all stars were assumed to lie. But new telescopes of unprecedented power and advances in photography revealed a universe unimaginably vast . . . and ten to twenty-five billion years old.

At the beginning of our century we saw the universe in terms of human experience. We took for granted the laws of gravity, the linear movement of time, the instantaneousness of sight and light, and the centrality of our solar system in a universe no larger than the cheerful night-light of the Milky Way. This was the Newtonian view of the universe, the popularization of the discoveries and theories of the great seventeenth-century British physicist Isaac Newton.

Throughout the world there were no better explanations for the birth of the universe than the local culture myths. Nearly everyone believed that what we could see in the bowl of heaven was a fixed reality and that what we see is all there is, as it had been since humankind first began trying to understand the patterns of movement across the skies and seasons.

But by 1900 rudimentary telescopes were revealing carpets and swirls of stars never before seen or imagined. Hundreds of thousands were listed and divided into groups by size and brightness. There were also unsettling new observations, including odd light smears and comets. How much more was hidden beyond the range of naked human vision, and what might it all mean?

By 1930 three revolutionary discoveries were to shatter our traditional views of our universe, the nature of basic matter, and the reality of time and space. But these revelations had their beginnings much earlier.

The Unruly Eye of Science

The American astronomer Percival Lowell's fascination with Mars in the 1880s inspired a generation of fine astronomers, though his own conclusions have not stood the test of time. Lowell, an affluent businessman and dedicated amateur student of the heavens, built his own observatory deep within the pine forest in the hills above Flagstaff, Arizona. Along with his assistants, he worked for fifteen years to satisfy his obsession with ferreting out the "secrets of Mars." Night after night, sitting beneath a wooden dome and peering through his telescope at the surface of the red planet, he began to see what his expectations agreed to see: evidence of humanlike achievements in the form of irrigation "canals" scratched out on the face of the planet.

Eventually he produced drawings of these supposed artifacts of ancient engineering. They stunned the world, stirred debate, and inspired shelves of imaginative fiction about the lost civilizations of our nearest planetary neighbor. Combining the observations of scientists around the globe, Lowell charted hundreds of the canals carefully. The

A page from the gentlemanly astronomer Percival Lowell's 1894 logbook shows how meticulously he recorded his conviction that the faintly observed markings on the surface of Mars were the remains of a system of irrigation canals dug by intelligent beings in antiquity. So dedicated to unlocking the secrets of the solar system that he built his own observatory in Flagstaff, Arizona, Lowell accurately predicted the existence of Pluto, but was not able to detect it before his death in 1916.

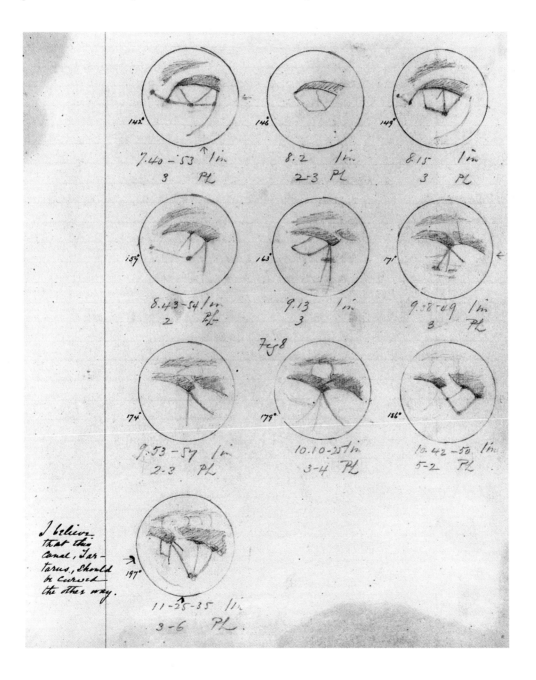

canals were considered proof of a great civilization's doomed attempt to ferry water from icy polar caps to cities dying from thirst as the ancient planet dried up.

But most professional astronomers remained skeptical. The telescopes of Lowell's day were not technologically sophisticated enough to produce a clear image of the surface of Mars. Nor did the camera, a relatively recent technological advance, provide much help: The film was grainy, and the required lengthy exposures tended to increase atmospheric distortion. The hazy images that resulted were not very satisfying. The "canals" were seen most clearly by unaided eyes and prepared minds. Soon, they faded into fiction. A half-century later, when the *Mariner 9* space probe orbited Mars in 1971, no canals and very few linear features could be found. And no evidence of Martian irrigation science has been revealed by the more than one hundred thousand close-up photos taken since.

Nonetheless, Lowell contributed to our search for knowledge, if only by intensifying our interest in his subject. Like all astronomers and physicists who explore the largest and smallest worlds of the twentieth century, he longed to learn the truth about the nature of the universe—its timeless secrets and its fundamental rules. He excited the public imagination. His vision fascinated the children who became this century's giants of astronomy and physics.

We Look Farther and Deeper

In 1897 a critically important device for piercing the heavens was constructed beside Lake Michigan at the Yerkes Observatory in Wisconsin. At age twenty-three George Ellery Hale, who was to become the century's most eminent builder of telescopes, had designed a lens forty inches wide, the first of three times he created a telescope larger than any other in the world when it was built. It took five years to construct the device, which was a refracting telescope. In this type of instrument, the observed light is refracted, or bent, as it passes through a convex lens, thus being focused on a point behind the lens.

As a child Hale was indulged in his endless scientific curiosity by a rich father, a Chicago elevator manufacturer who amassed a fortune when the city rebuilt after the

The very driven George Hale, reclining on the ground, seems uncharacteristically relaxed in a glade near Southern California's Mount Wilson Observatory, where he designed and supervised the construction of the most important telescopes of the first half of the twentieth century. His companion, Ferdinand Ellerman, became Hale's first assistant there in 1904, and was invaluable for his extraordinary skills as mechanic and staff photographer.

devastating fire of 1872. At age thirteen, the boy drew upon the family resources to build a laboratory for himself and his two younger siblings. It was equipped with a metal lathe, steam engine, telescope, microscope, and spectroscope.

Hale was intellectually adventuresome and thoroughly transfixed by science, as he recalled later:

> *Each of us had a seat and an "outfit" consisting of [a] Bunsen burner, batteries, galvanometers, and other devices. We poured hydrochloric acid on zinc and lit the evolved hydrogen. We mixed potassium chlorate and manganese oxide and collected the oxygen set free. We decomposed water into hydrogen and oxygen and reunited the gases by a spark. Often our delights were enhanced by frightening but delicious explosions. . . .*

Hale's enthusiasm led him to the Massachusetts Institute of Technology, where he became focused on the mysteries of stellar phenomena. What was most needed to advance knowledge of the heavens, he decided, was improvement in telescope design. To that end the further need was for deep pockets, since the U.S. government was not yet given to in-

vesting in expensive excursions into pure science. Luckily young Hale had a remarkable talent for prising funds loose from nineteenth-century moguls—not only his own father but also a Chicago streetcar tycoon, a hardware king in Los Angeles, and the legendary steel magnate Andrew Carnegie. Their money made possible his famous telescopes.

Hale's second effort was a sixty-inch-wide reflecting telescope set up in 1917 on Mount Wilson, an isolated peak near Pasadena, California. This type of telescope, also known as a Newtonian telescope for its famous inventor, gathers more light than a refractor and can easily be built larger. The light from a celestial object is reflected by a concave mirror to a focal point above the mirror, then beamed to the astronomer's eye by a flat mirror or secondary convex mirror. The larger the concave mirror, the more light is collected, thus the farther the visual probe into outer space.

Shown from the south, the huge shutter assembly in the dome of Mount Wilson Observatory slides open to allow light to hit the hundred-inch telescope, the world's largest from 1917 to 1948. Astronomers worked all night, even when exposed to the frigid air of winter at an altitude of five thousand feet, in order to fix an image of starlight on a photographic plate. One of these photos led to the astonishing discovery that the universe is expanding.

During the fourteen years it took to construct this ambitious telescope, it became increasingly clear that Hale was one of science's doomed heroes. His tenacity and inability to resist daunting challenges were somehow linked with the weaknesses of his psyche. Each achievement, though acclaimed, wore away at him. Nervously enduring the long waits necessary with each daring new design, Hale became convinced that he suffered from "Americanitis," a national disease in which the ambitions of Americans drive them insane. He had headaches, sleeplessness, indigestion, and tingling feet.

But he plunged ahead. He realized that his telescopes so far revealed only more stellar objects to catalog, not breathtaking new types of discovery. Even before his second telescope was fully tested, he began building in 1908 a reflector with a hundred-inch-wide primary mirror. Perhaps it could raise the science of astronomy to an entirely new level of observation, pushing theory past the old limitations. This project, which cost millions, added ten years of the double stress of protracted construction and deferred observation.

The risks taken were enormous. To cast the largest solid piece of glass ever made until then, a French specialist poured the molten equivalent of ten thousand champagne bottles into a huge mold packed with heat-retaining manure to allow the glass to cool slowly. The first effort was flawed, the second and third were cracked, and the fourth was even more imperfect than the first. At last, hoping the flaw would not prove fatal, Hale decided to send the first attempt to his opticians. It would be impossible to know the outcome of this decision for certain until years into the project.

For the body of the telescope and the dome to house it, steel and ironwork were fabricated in a shipyard on the East Coast. Too large for boxcars, the steel sections were shipped around South America to California and hauled up to the summit of Mount Wilson. Special roads had to be built. The iron and steel of the telescope weighed 100 tons, while the dome weighed some 650 tons. The support site became a kind of technological kingdom with machine shops, living quarters, and research facilities.

The telescope was ready for its first use on November 2, 1917. Hale, tense as always, was horrified to focus this monstrous device on the largest planet, Jupiter, and find multiple images. Steeling himself, he decided to believe that the mirror had not cooled down in the air of evening. He and his assistant waited through the long, cold, silent California night. Finally there emerged only one Jupiter, luminous and beautifully distinct. The gambles had paid off.

One of the most notable engineering wonders of our century, the hundred-ton, hundred-inch reflector moved as precisely as a fine watch, using a revolutionary mercury flotation system and thirty-five coordinated electric motors. Even as the earth spun beneath it, the telescope focused without wavering on a specific spot in the heavens. The nine-thousand-pound primary mirror, perfect after grinding to one part in ninety-two thousand, could detect the flickering of a candle at five thousand miles. More to the point, it probed beyond the Milky Way, gathering light no human had ever seen before.

One of the astronomers recalled the sensation of working with the remarkable new behemoth: "You were alone on the mountain with a telescope. It was just you and the universe, or you and God so to speak, and the instrument."

Unfortunately, by this time Hale was no longer the focused rationalist experimenter or the elated enthusiast. Pushed finally to the edge by the decades of ongoing stress, he suffered a nervous breakdown and made the first of numerous visits to a sanatorium for rest. From this first episode onward, to the distress of his admirers but per-

haps as a way of dealing with his strong drive, he preferred to discuss both his telescope and life in general with his new companion, a sympathetic green elf or leprechaun. Still, he would be able to produce even greater achievements for the future of astronomy and physics—including a two-hundred-inch telescope—but his dedication took an exhausting personal toll.

The Monastery of Astrophysics

The daily routine was extremely demanding at the isolated Mount Wilson Observatory, which could be reached only by a half day's strenuous hike or by mule train. It did not offer much in the way of escape or diversion. Still, more than a dozen astronomers came to work at Hale's new site in the early years. They stolidly sat out all night on the observatory platforms, even in the deep cold of winter. For up to twelve hours they stared at the illuminated crosshairs of the telescope sight, fine-tuning the magnificent Hale reflector with a small hand paddle. At times their tears froze to the eyepiece.

While they oversaw the sustained photographic exposures necessary for revealing phenomena deep within outer space, they could not even warm themselves with coffee since Hale believed the beverage was physically harmful. Some wore coats made from bearskin or sheepskin. Nor did they know if their ordeal was worthwhile until the next day, when the glass photographic plates were developed.

Relations between staffers were numbingly formal; the atmosphere was nearly as airless as space itself. Among other restrictions imposed in this scientific cloister, women were not allowed to stay in the dormitory, inevitably known as the monastery. Nor could female visitors even touch the costly hundred-inch telescope, much less sign up for viewing time. Hale believed that astrophysics demanded a masculine devotion free of all distractions.

Such a barrackslike, frontiers-of-science setting was perfectly in tune with Edwin Hubble's dramatic appearance on Mount Wilson in 1920. Wearing knickers, by then an antique fashion, and ornamented with dueling scars that others claimed he had inflicted

Workmen carefully polish the solid-glass hundred-inch mirror that was the heart of Hale's magnificent telescope. In a reflecting telescope like Hale's, the surface of the mirror is curved and very highly polished to focus the celestial image with a precision remarkably close to the wavelength of light. This image is then deflected by a much smaller mirror to a magnifying eyepiece for a human observer or to a photographic plate.

on himself, this native of Marshfield, Missouri, spoke with a British accent, an affectation contracted when he was a Rhodes scholar studying law at Oxford University. But if he was perhaps a poseur, Hubble was by no means an impostor. As a boy he learned to love astronomy by sharing his grandfather's telescope, as well as the old man's fascination with Percival Lowell's fantasies about Mars. Driven to succeed, Hubble had become a star boxer and basketball player in college and an army major in the recent world war. He enjoyed being seen as separate from other human beings.

If his colleagues found his pretensions insufferable, they also recognized with something like awe that he was inimitable on more than one level. As he scanned the skies, pipe glowing in his mouth, his extraordinarily disciplined, eager, wide-ranging intelligence continually processed what was seen into what could be known or guessed. To others, it seemed that the act of seeing for him was almost synonymously an act of conceptualizing. Moreover, according to his envious coworkers, the cosmically analytical Hubble could control his bladder for extremely long stretches of time—no minor gift in his line of work.

Hubble had both a flair for asking the right scientific question and an abiding curiosity that made him famous. He was obsessed with the tens of thousands of fuzzy patches of light in space known as nebulae, from the Latin word meaning "mist" or "vapor." These unexplained phenomena had been the subject of his Ph.D. dissertation at the Yerkes Observatory. Other scientists had dismissed this area of inquiry as a distracting anomaly. Why chase after this indistinct luminescence when bright stars and fabled planets, comets and constellations, seemed rich with unprobed information?

Hubble steadfastly worked at the Hale hundred-inch telescope for four long years, taking thousands of photographs of nebulae on his personal quest. Then he discovered what no one had ever known before: These pale clouds of light are in fact mobs of unknown, uncountable stars, intimations of galaxies greater in number than anyone had ever seriously imagined. His photographic plates became actual black-and-white proof that the

The astronomer Francis G. Pease, one of the few allowed to work with or even touch Hale's delicately tuned instrument, uses the telescope's eyepiece to measure the distance of a star. Partially visible at right is a portion of the gigantic machinery designed to move the heavy mirror gently and in perfect synchrony with the earth's motion, so that the telescope remained fixed on a specific object or span of sky.

stars of the nebulae lie *outside* the Milky Way—that is, outside what was then the known universe.

On an October night in 1923 he took the photograph of his life. Focusing on the great nebula known as Andromeda, he made a forty-minute exposure that showed a nova. At the time the term referred to a star that unexpectedly—and inexplicably—suddenly grew brighter. The following night, when the air was clearer, he believed he captured three novas. But out of the developing bath came an image even more astonishing, one of the strange celestial objects known as Cepheid variables. The excited Hubble wrote "VAR!" on the slide; the letters mean "variable star," and the exclamation point means "eureka!"

The enormous significance of Cepheid variables had been plumbed by someone unable to work alongside Hubble, Henrietta Swan Leavitt, one of the rare female astronomers of the day. At the time the Harvard College Observatory hired women to work in the field, but only as scanners of photographs collected by male astronomers.

An inspired researcher undeterred either by her deafness or by the monotony of systematically poring over piles of photograph plates throughout her entire career, Leavitt specialized in comparing the relative brightness of celestial objects. This task earned female researchers the name of computers—and for good reason. Typically the analyst might examine images of ten thousand stars again and again in order to identify a single one that varied in brightness.

Leavitt had become especially expert in the characteristics of these variable stars, whose luminosity dims and flares up in cycles, and eventually cataloged about twenty-four hundred of them. Most important for Hubble's research, she was able to time the cycles of the Cepheids. By showing that the length of their cycles of luminosity varies in close relation to their brightness, Leavitt provided astronomers with a hitherto unknown celestial yardstick, a method of measuring the distance of stellar objects. If a nearby Cepheid with a certain cycle has a certain degree of luminosity, a remote Cepheid with the same cycle should have the same luminosity. The degree of its relative dimness, from our point of view on earth, shows how far away it is.

Leavitt understood precisely the theoretical importance of her discovery as early as 1912, but her supervisor hustled her back to collecting data. Hubble, on the other hand, used Leavitt's yardstick in 1929 to calculate the distance to the Andromeda nebula, cataloged as M31, which was thought to be no more than 100,000 light-years away—well

Affecting the sporting attire of a British swell, but born in the Midwest, the American astronomer Edwin Hubble, shown here at the Mount Wilson Observatory in 1923, was actually a prodigiously talented, tenacious researcher. According to colleagues, he could work longer hours than anyone else on the mountain. With the help of his volunteer assistant M. Humason, a mule driver on the observatory's supply train, Hubble compiled mounds of data on the redshift of galaxies.

within our own Milky Way. To Hubble's astonishment, it turned out to be some 930,000 light-years off in space:

> *We can write the figures [he wrote], but they are utterly beyond our comprehension. When we look into the depths of space, we are gazing back into history. The nearest star, we see today as it was four-and-a-half years ago; the nearest nebulae, as they were 100,000 years ago; the frontiers of the known universe as they were 200 million years ago back in the carboniferous age of the geologists. [NOTE: Today, Hubble's original measurement of 930,000 light-years has been recalibrated to 2,200,000 light-years.]*

Andromeda and indeed all the other nebulae across the heavens are galaxies containing vast numbers of stars and lie vast distances away. With Hubble's work our own solar system became verifiably small and insignificant—at least in terms of physical size and distance and evident uniqueness. Our life-giving sun is a speck of sand on the cosmic beach.

Working at the Harvard College Observatory at the turn of the century, Henrietta Leavitt developed the information about Cepheid stars that would make Edwin Hubble's great discoveries possible. Noticing that these stars varied in brightness at a consistent rate that was slower for the brightest ones, she recognized that they could be used to measure stellar distances. Applying this celestial yardstick herself, Leavitt showed in 1912 that such variable stars had to be far outside our home galaxy.

The vigorous jotting in the upper right-hand corner—"VAR!"—shows Edwin Hubble's delight upon analyzing a photo taken with the Hale telescope during the night of October 6, 1923. It refers to a Cepheid variable. The one marked in this famous photo proved that the Andromeda nebula is a huge galaxy enormously distant from the earth and completely independent of the Milky Way.

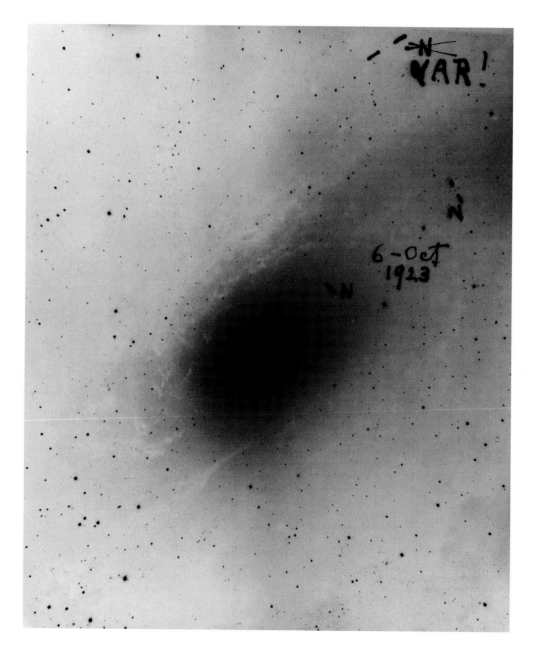

A universe that had been defined for aeons by what the unaided human eye could see, if not always clearly, was perceived after Hubble's breakthrough as a universe that might well be infinite, galaxies beyond galaxies without boundary. The scientific community and indeed the rest of humankind had to deal with a discovery that is fundamentally a revelation—thanks to Hale's invention, Leavitt's discovery, and Hubble's tenacity.

Then Hubble turned his unusually original mind to an inspired question. For six intense years he and his meticulous technician, Milton Humason, worked long nights amassing data about the distances of galaxies. Is there, he wondered, a connection between these distances and the movements of the stars?

Scientists knew that galaxies are not fixed in place; they are all dancing through the heavens. If one is moving away from us, the light reaching the earth is stretched to the longer wavelengths of the red end of the visible spectrum, or redshifted. The greater the redshift, the faster the velocity of the fleeing celestial object. When a star is moving nearer, its light shifts toward the blue end of the spectrum.

By focusing on the far galaxies discernible only to the hundred-inch Hale telescope, Hubble made an astonishing discovery, arguably the most significant for science in this century: The farther the galaxy, the faster its flight from us. In other words, its ve-

locity is directly proportional to its distance. This insight became known as Hubble's law, and its implications were both exciting and troubling when it was announced in 1929, for it means that the universe is continually expanding—its galaxies racing ever farther apart—and not unchanging.

One physicist recently summed up the immediate, disturbing implications: "How could that be when the whole history of human thought assumed that the universe is this fixed thing? And since the universe is everything, how could it have an evolutionary path? An expanding universe might have to have a beginning . . . and an end."

Perhaps since the first humans stood up to walk the earth, we have known that human affairs are unstable. In the early twentieth century we began to learn that the heavens themselves are unstable.

Many physicists now joined with astronomers in trying to make a new kind of sense of Hubble's discovery and other breakthroughs in astronomical observation, but they had been tussling with odd and unsettling revelations of their own. In the years since the turn of the century, the accepted rules of this branch of science began to crumble or distort themselves, it seems, as experimenters tried to grasp the meaning of unforeseen experimental results.

The Patent Clerk

In the 1890s scientist and layperson alike perceived the physical world through Isaac Newton's eyes, more or less. If his explanations of planetary movement were not widely or fully understood, most people did believe that his work meant that gravity causes objects to fall to earth, that time is a river moving at a constant speed toward the future, that solids are solid, and the distance of a mile is always a mile. The world ticked along like a huge mechanical clock. Newton's insights predicted exactly how a pendulum would swing or a planet glide through the heavens.

At the same time the tenets of Newtonian or classical physics had never been so widely accepted by the public. Anyone could see its enormous impact upon daily life: streets being dug up to lay down electrical cables to light up city homes and businesses,

In an age when very few women were given opportunities in science, the women of the Harvard College Observatory, shown here in 1925, were restricted to classifying celestial phenomena in photographic plates taken by male astronomers. Antonia C. Maury, third from right, like Leavitt, was an original thinker. She classified stars by the width of the lines they produced on a spectrograph, a tool that would help others nail down the exact dimensions of the Milky Way.

dynamos being built to convert mechanical energy into electrical energy by rotating conductors in a magnetic field and thus providing vast reserves of electric power. When professors of physics, often in the public lecture hall, scribbled on a chalkboard, everything seemed in control, for their equations explained the predictable workings of the phenomena that fueled progress: electricity, magnetism, heat, and light.

But there were unexplained gaps in this carefully reasoned world. The work of Albert Einstein, among others, was part of the flood of new discoveries that swept away that classical physics forever.

An obscure German technical clerk, third class, at the Patent Office in Bern, Switzerland, Einstein graduated from the Swiss Federal Polytechnic in Zurich but was unable to find an academic job in physics anywhere in Europe. Nearly impoverished, he was hired

by the Patent Office in 1902. He married, started a family, but had only one passion, the mysteries of physics. Fortunately his job made use of and probably helped direct his venturing intelligence. Einstein's assignment was to strip each patent application down to the bare bones, hacking through the verbiage and illustrations to determine whether or not a new idea really did lie within. For this particular genius, who became the most influential agent of change in modern physics, this method proved to be an instrument of unique clarity. He was to use the same approach in the arcane regions of theoretical physics.

Einstein once explained the clarity of his work this way: "Children who wonder about things like light, time and space are satisfied with stock answers and never give them another thought as adults. But because I was a late developer, I first pondered such 'simple' questions as an adult, and so probed them more deeply and tenaciously than any child would do."

From 1895, when he was sixteen years old, Einstein was especially puzzled by the problems raised by the behavior of light, a mystery to be solved in the mind, not in the laboratory. He eventually spent years devising thought experiments to answer speculative questions that absorbed him almost constantly. Even as a schoolboy Einstein knew that experiments proved that light is a wave or that it moves something like any wave through any body of water. He also knew that the physicist James Clerk Maxwell had shown that light moves at 186,000 miles per second.

In his first important thought experiment, the sixteen-year-old happened upon a stubbornly resistant problem. According to Newton's physics, we can catch up to a wave and match its speed of movement. In that case the wave will seem frozen since observer and wave are traveling at the speed. But Einstein was convinced that light would not be light if you could catch up to it. It would be impossible to "see" light not moving, since vision requires the movement of light from object to eye. Einstein could not find his answer for ten years. He would not let the problem go, but traditional physics courses at high school and his college years in Zurich provided little help. On the other hand, he often skipped school to read or wonder on his own.

At age thirteen or fourteen, Albert Einstein poses before a sentimental backdrop in a photographer's studio. Einstein later recalled that he had been fascinated by the problem of light since childhood. The currents in the water suggested by the backdrop, the relative motion of the sailboats in different parts of the bay, and the movement of light from camera to face and back again were the kinds of images he pondered until he came to conclusions that revolutionized the way human beings understand time and motion and space.

In 1905, in the midst of a stroll with a friend, the adult Einstein abruptly broke off a casual conversation mid-sentence and raced back to his two-room third-floor apartment. Somewhere on the back stairs of his brain he had just put together the critical equations of the world-shaking special theory of relativity, perhaps the century's most important mathematical operations, and needed to write them down. This was after all the culmination of the work of a decade.

What Einstein began to understand about the true nature of physical reality may seem counterintuitive, but his illustrations are as clear as the famous illustration of Newtonian gravity, the apple supposedly falling on the theorist's head. Einstein's thought experiments should be read for what they show, not what they mean. (After all, no one really understands gravity, the convenient apple notwithstanding.) Einstein's homely examples reveal how time and space interact, not why they do so. They are also pure fun, for this was a physicist who really believed he was at play in the fields of the Lord.

First, Einstein assumed that the speed of light is constant, even when measured by people racing past each other at constant speed. Instinctively that just made sense to him. With equal self-confidence he also assumed that there is no difference in the laws of physics in two frames of reference moving at constant speed with respect to each other. These two assumptions are easily understood in his thought experiment of the passenger holding up a mirror on a train moving at the speed of light.

What does the passenger see when he looks in the mirror? Possibly there is no reflection because the speed of the light traveling toward the mirror and the velocity of the train are the same. No, Einstein said as early as age sixteen. That doesn't and can't make sense; the passenger must be able to see his reflection.

But if the passenger does see a reflection, what about someone standing beside the railroad tracks? This observer sees the train moving at the speed of light. Does that mean the light traveling to the mirror must move at twice that speed, or 372,000 mps, for the person on the train to see his face reflected? That would be the conclusion of classical physics. But Einstein said that doesn't and can't make sense either. The speed of light must be the same always, no matter whether it's measured on the train or from beside the tracks.

How, then, can the train passenger and the stationary observer have their apparently contradictory experiences?

To solve this and similar thought experiments that would bewilder others for decades, Einstein had gone back to basics in his decade of thought experimentation.

Speed, by definition, is distance divided by time. Einstein realized that distance and time must not be absolute, or fixed, as they were in the world revealed by Newton. Instead they are relative. In other words, if an observer on the railroad platform and an observer on the moving train each holds up metersticks and clocks, the person standing on the platform will find that the moving meterstick is shorter than a full meter and the moving clock takes longer than one second between ticks. But that's not the whole story, for the passenger on the train will find that the meterstick and clock on the stationary platform are short and slow by exactly the same amounts!

The effects described in Einstein's paper on the special theory of relativity become noticeable, however, only for speeds near the speed of light. A spaceship traveling that fast might seem from the earth to take fifty thousand years to reach the edge of our galaxy while astronauts on board experience only two weeks of time. The time frames on earth and in the spacecraft are subjective. Put another way, the light from clocks on the earth would take so long to catch up with the vehicle that the tens of thousands of earth years would seem to be mere days.

In addition, everything gets heavier near the speed of light. Physicists speculate that actually revving up a spaceship to half the speed of light and stopping it would require fuel almost seven thousand times the mass of the craft itself. All the fuel currently available on the planet would not be enough to push a spacecraft near the speed of light. Besides, the distance to a galaxy in light-years is also the number of years it would take to reach there, for the speed of light is a kind of cosmic speed limit. In other words, Einstein's special theory of relativity seems to suggest that intergalactic or interstellar space travel is just not possible.

But Einstein was dissatisfied because his theory was not inclusive enough. It applied to separate reference-frames that move with constant velocity in relation to one another; he wanted to include all movement everywhere. For example, if a windowless elevator is accelerating upward in deep space and a laser beam is fired upon it, what will a passenger in this elevator see? The beam will enter one wall at a certain height and, as the elevator continues racing upward, exit the opposite wall at a lower height. In other words, to the observer inside, the acceleration has somehow caused the light to travel in a curved path between the two walls.

This seems impossible according to the laws of classical physics. The light must travel the shortest distance between two points, and that is a straight line. Einstein made another extraordinary intuitive leap, which he called "the happiest thought of my life."

Assuming that acceleration and gravity are the same phenomenon, he saw that gravity must be able to bend light. The shortest distance between two points is a curve if the points lie on a curved surface. Therefore, light would travel in a curve if space and time, like physical objects, were curved.

After years of arduous mathematical calculation, he showed that space and time are indeed curved. Eventually he proved that they are curved by the effects of mass. In 1915 he presented these conclusions to the world in the general theory of relativity.

Einstein's new theory of gravitation was barely noticed at first, in part because most physicists of the day were not familiar with the mathematical techniques he used to explain space-time. But in 1919 an expedition of British scientists made an observation that brought immediate international attention to the man and his discovery. During a solar eclipse, they measured the bend in starlight passing the massive, darkened sun. When they compared this bend with the path taken by starlight that did not pass near the sun, they found that the difference between the two pathways matched well with Einstein's predictions. In sum, the observation showed that light really does slide along a curved space-time.

The witty Irish playwright George Bernard Shaw paid tribute to Einstein and his relativity theories with tongue in cheek: "Ptolemy created a universe that lasted a thousand years. Copernicus created a universe that lasted four hundred years. Einstein has created a universe ... and I can't tell you how long it will last!"

THE REVOLUTION IN SCIENCE/EINSTEIN VERSUS NEWTON, exclaimed a headline in *The Times* of London on November 8, 1919. From then on, much to the amazement of this obsessed loner who had trouble sustaining human relationships of any depth, he was a cultural symbol. His arcane theory in mathematical physics was thought to hold the key to understanding the nature of the universe. He had proved, to the distress of many, that physical concepts are, in his words, "free inventions of the human intellect."

"The Most Incredible Event"

In contrast with Einstein's solo, the other critical advance in theoretical physics in the century's first decades—just as shocking as the rarefied concepts of special and general relativity—was the product of a group effort.

Emerging alongside the new picture of the universe—billions of stars, billions of galaxies, moving in warped space and time—came the theory of quantum mechanics. It eventually described a world that made no sense to anyone, including the brilliant physicists who discovered its principles, yet the logic of equations would show that the inconceivable is normal at the subatomic level. Deep in the heart of things, particles don't exist unless someone looks at them, the path of a particle from *a* to *b* cannot be plotted, particles may communicate with each other faster than the speed of light, and the fundamental rules of matter are chance and uncertainty.

In the Cavendish Laboratory at Britain's Cambridge University, Newton's academic home for his entire career, a young scientist from New Zealand faced an ultima-

Below left: A New Zealand native who delayed his marriage in order to accept a scholarship at Cambridge University, Ernest Rutherford would eventually achieve worldwide fame, a Nobel Prize, and marriage to the love of his life. Among his many achievements, the most influential was his work on the structure of the atom. Below right: In 1911, he discovered that every atom has a nucleus with three unexpected and previously unimaginable characteristics: It carries a positive electrical charge, represents virtually all of the atom's mass, yet is smaller than 10^{-14} of the atom's volume. The electrons, which have virtually no mass, have to orbit the nucleus at distances that are comparatively great in subatomic terms. To explain this revelation in terms of Newtonian physics, Rutherford created a dramatic new picture of the atom. This drawing conveys the basic idea but not at all the relative distances and sizes. The electron zipping around farthest from the center would have an orbit larger in diameter than the nucleus by something like a hundred thousand times.

tum at the turn of the century straight out of melodrama. Ernest Rutherford longed to marry his sweetheart back home, and his plan was to study radio waves and make a commercial success. Yet the great scientist J. J. Thomson, director of the Cavendish, was implacable: Decent folk chose between the high road of pure science and the decidedly low road of making mere money. In 1897 Thomson had discovered the electron, a negatively charged particle found in all atoms, and was interested in plumbing even deeper into the nature of matter.

Rutherford elected the side of the angels—in this instance, the pure science of seeking an explanation for radioactivity and X rays. After all, he had traveled halfway around the world to learn from a mentor, as was the custom of the day in science. Unsophisticated and disturbed by the bared arms of fashionable ladies of elite Cambridge society, he threw himself into his work.

By chance or intuition Thomson had chosen well, for young Rutherford revealed a special knack for the field. Between 1899 and 1901 he discovered the concept of radioactive half-life, the period of time in which a material decays enough to lose half its radioactivity. This radioactive decay, he recognized, is a natural process in which one chemical element changes into another, an indication, perhaps, that different chemical elements are made from similar basic materials. Rutherford also found that radioactive emissions have at least two forms: alpha rays and beta rays. (A third form, gamma rays, was discovered later.)

Thus proving himself, he married his sweetheart and became a full professor at Manchester University. It was in a basement laboratory there, where an experiment could be performed in near darkness, that he achieved one of the century's landmarks in physics.

Based largely upon Thomson's work in the 1890s, the shared wisdom about the smallest known unit of matter, the atom, used the imagery of a familiar unit of daily British life, raisin pudding. The atom was seen, in other words, as a solid but permeable, diluted mass with electrons frugally dotted about within it. The whole soggy mass was held together with a positive electric charge.

Alpha particles, which are ordinary helium nuclei, stream out from radioactive sources at some ten thousand kilometers per second. These tiny flying bullets can be used to detect the structure of atoms: Where they change their path slightly, they must have met a subatomic obstacle. In 1910, when Rutherford and two assistants trained a beam of alpha particles upon a screen made from very thin gold foil, he naturally expected all the alpha particles to pass through the atomic pudding, even if a few were knocked off course. Not so. Occasionally one particle in ten thousand bounced back.

He was dumbfounded. Not even one in a million should ricochet back, since he knew that the alpha particle weighed eight thousand times as much as any electron in the "pudding" and was traveling at a velocity of ten thousand miles per second.

"It was quite the most incredible event that has happened to me in my life," Rutherford recalled later. "It was like a 15-inch artillery shell hitting a piece of tissue paper and bouncing back."

The explanation was in no way obvious, and Rutherford spent a year struggling against the traditional enemy of scientific advance, the preconceived notion. Eventually he grasped that the atoms of the gold foil would be sturdy enough to deflect alpha particles only if all the mass of the atom was gathered together in a small, dense nucleus with a concentrated positive charge, with the negatively charged electrons surrounding it. In short, he made a stunning leap in understanding: The atom is almost entirely empty space, whether it is an atom of satin or steel, Kool-Aid or concrete. Most of matter is a void. The atom has a tiny center with a positive electric charge. Its electrons, which are negatively charged, orbit about this nucleus like the planets circling our sun.

Like Einstein's speculations, this is another discovery that seems to defy common sense and certainly classical physics. To adopt an analogy used by the physicist Malcolm Longair, who works at Cavendish today, we can picture the atom as the size of a soccer stadium and its atomic nucleus as the size of a soccer ball. Imagine the ball in the center of the field. Then its orbiting electrons would be whizzing around the top row of the stands. But as Longair points out, the essence of Rutherford's discovery lies in the nature of that fictitious ball: It would be so dense that it would contain nearly all of the stadium's mass.

Why, then, do primarily void soccer balls not pass through primarily void stadium walls in real life? Because each atom has an electric field. The field around the charged particles in the ball repels the field around the charged particles in the bricks and mortar.

On May 7, 1911, Rutherford explained his amazing finding to the Manchester Philosophical Society, where he chose to make his first report. He shared the bill with a fruit wholesaler who had discovered a rare snake in a recent shipment of imported bananas. To the members of the society it was apparently just another meeting made diverting by the endless series of amiable wonders that alert gentlemen might encounter. Neither discovery, atomic structure or reptile, made a stir.

Professional physicists noted Rutherford's claim but dismissed it for violating the laws of classical physics. The electrons orbiting around the positive nucleus would lose energy as they continually gave off radiation. Logically they should spiral into the nucleus within a fraction of a second. In short, Rutherford's imagined atom would be unstable. The universe would collapse into atomic rubble. Matter is matter, not vast, electrically charged, virtually empty space. Rutherford lost faith in his own model of the atom.

"A Little Information . . ."

But his great discovery would be saved, if strangely modified, by the bold insight of one of his students, an athletic young Dane named Niels Bohr. After working under Rutherford for only a few weeks in 1912, he wrote his brother a tentative letter that unofficially an-

Niels Bohr not only explained what was wrong with Rutherford's view of the atom, he discovered a physical reality so mysterious and apparently lawless that scientists angrily resisted it at first. He saw that classical physics would never be able to explain why the weak electrons do not fall into the powerful nucleus, first causing the atom to collapse, then causing the entire universe to collapse. His answer was the so-called quantum jump: Atoms are stable because the electrons can exist only in specified circular orbits and can change from one orbit to another only by leaping in an instant.

nounced a concept that would startle, annoy, and divide the physics community: "Perhaps I have found out a little about the structure of atoms. Don't talk about it to anybody. . . . It has grown out of a little information I got from the absorption of alpha rays."

His breakthrough became the century's second concept-shattering theory in physics, and it is even more difficult to conceptualize and believe plausible than Einstein's ideas about relativity.

Bohr's solution owed much to the thinking of the German physicist Max Planck, who had proposed that matter and radiation are not absolutely continuous. Because of his interest in thermodynamics, the study of heat, Planck concluded that the energy in all electromagnetic waves exists in the form of tiny, discrete amounts, or quanta. In classical physics, all changes in energy levels were pictured as a continuous curve. In much the same way, light was also conceptualized as a continuous wave throughout the nineteenth century. But in 1900 Planck published his conclusions, which applied to heat, light, and radio waves. In 1905 Einstein seconded this insight, affirming that light could be understood as a collection of discrete quanta later dubbed "photons."

Bohr proposed his quantum model of the atom in 1913, but the atomic structure of all elements could not be explained this way. Eight years later, he drew this diagram in his research notes to describe an important revision: Elliptical orbits could be used as a model for all elements. This was the first comprehensive, all-inclusive theory of atomic structure. Still further revisions would be necessary, but his work was judged accurate enough by his peers to earn him the Nobel Prize for Physics in 1922.

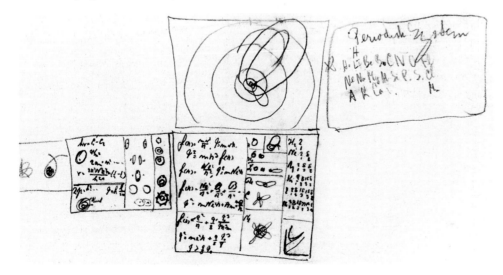

Bohr latched on to this idea that energy and light are made of tiny, unchanging particles and thereby explained Rutherford's atom. The orbits of the electrons, he proposed, are another example of discontinuity in nature. The electron cannot move, as classical physics would dictate, in a continuous line toward the nucleus. Only certain orbits exist, each a different energy state that is clearly defined. Electrons have no choice but to leap between them. In addition, the electrons radiate energy only when they make the jump from one orbit to another.

It is essential to this view that absolutely nothing exists between the defined orbits. Traditional physicists found this notion nonsensical. Could you not, for example, move the earth's path of revolution just a bit closer to the sun?

The wise men of the academy accused Bohr and his adherents of relying upon fantastic assumptions. Said one: "This is just a cheap excuse for not knowing what's going on."

Yet Bohr's model solved one of the knottiest problems of the previous century's physics. With a spectroscope it is possible to determine a unique spectrum—or series of wavelengths of light or radiation—for each element. Why should each element have its own spectrographic signature? Bohr had just solved the mystery. As the electron leaps to a higher orbit, the spectroscope sees a dark line; if it leaps downward, it produces a bright line. These so-called quantum leaps identify the characteristic atomic structures of all elements, the number and spacing of electrons around the nucleus.

To Einstein, this gigantic leap in conceptualizing the submicroscopic world of the atom was "the highest form of musicality in the sphere of thought." It was also, despite the modifications and corrections that became necessary, the inspiration for the work of a small close-knit generation of international theoretical physicists. For most, the place to be was the Institute for Theoretical Physics in Copenhagen, which Bohr set up with a grant from the Danish government in 1920. Eventually two views of atoms were defined there and theatrically advocated by eager young physicists screaming at one another.

In 1926, Erwin Schrödinger introduced wave mechanics by arguing that the position of an electron can be determined only when it is regarded as a wave. In the bizarre world of quantum mechanics, in other words, the particle's exact location must be indicated by equations that can describe only probabilities. His insight helped explain why light, for example, sometimes seems to be made of particles and sometimes seems to be made of electromagnetic waves.

In 1925 Werner Heisenberg, a twenty-three-year-old German, seemed to prove Bohr's picture of particles leaping from fixed orbit to fixed orbit with a series of complex equations. Bohr was relieved, but others found the math too difficult and the theory still inconceivable.

A year later Erwin Schrödinger, a young physicist in Austria, provided the alternative view: Atoms acted like waves, not particles, and that could explain the orbits of electrons. Because traditional scientists were familiar with the properties of waves from classical physics, Schrödinger's proposition was immediately popular within the trade.

But these were two radically different concepts. When particles meet with each other, they bounce off or smash or exert force. When waves meet, they can pass through each other or cancel each other out or produce larger waves. There was no obvious way to make these two theories meld.

And yet the strangeness became only stranger. In the summer of 1927, Heisenberg

pondered further consequences of his new quantum theory. To look at an electron, he realized, you have to shine a light on it. But light is a stream of photons that will knock the electron to a different position. In other words, as a tenet of the Heisenberg uncertainty principle, the fact of observing an object inevitably changes its location. It is therefore impossible ever to know where it really is. Moreover, the more you know about the particle's position, the less you are able to learn about its speed and direction, and vice versa; the more you know about a particle's speed and direction, the less you can know about where it is at any given time.

How did this insight, which is difficult to conceptualize in itself, help solve the particle/wave debate? Heisenberg saw that the electron is a particle whose *probable* position can be plotted as a wave. In other words, the probability of finding it at any specific point is a wave, and thus the wave and particle theories could be reconciled.

Once again common sense was rudely assaulted. Heisenberg's ideas directly contradict our experience of visible objects in the world we experience every day of our lives.

A puzzling example might help clarify the concept. If light is beamed in a lab experiment through two parallel slits toward a screen, two separate beams emerge from the slits and form on that screen an interference pattern of dark and light bands. According to classical physics, the pattern results because light is a wave and the two beams are reinforcing, neutralizing, or otherwise interfering with each other.

But if light is directed one photon at a time at the two slits—in other words, one particle at a time—the same interference pattern is produced. Why would the photons "choose" to be either dark or light? Why wouldn't they pass through the slit and arrive singly? No one knows. What path do they take from the source of light to the screen? It is impossible to tell. But it has become clear to researchers that a single photon can go through both screens at once—that is, be in two places at once. No one knows how that is or can be possible.

But we do know that formulas that describe these weird behaviors of the quantum world predict the behavior of atoms with incredible precision.

Still, even some of the greatest minds were uncomfortable with all this at first. Newton's mechanics aimed to predict the results of individual experiments with 100 percent accuracy. Quantum mechanics revealed that one hundred identical experiments at the subatomic level of physical existence will not each have predictable outcomes. An average result, but not the result of each individual experiment, can be anticipated. In the late 1920s science moved from a world of rules to a world of chance and uncertainty.

Einstein, whose work is partly responsible for Bohr's conclusions, was particularly

German theoretician Werner Heisenberg, shown looking puckish at about age twenty-four, is some two years away from startling the physics community—and at the same time resolving a knotty problem in conceptualizing the subatomic world—by introducing his famous principle of uncertainty: It is fundamentally impossible to know both a subatomic particle's exact speed of movement and its exact position at the same time. In effect, this insight marked the discovery of quantum mechanics.

skeptical: "Quantum mechanics is very worthy of regard, but an inner voice tells me that this is not the true Jacob. The theory yields a lot, but it hardly brings us any closer to the secret than the old one. In any case, I am convinced that He doesn't throw dice."

Bohr countered: "It's not our business to prescribe to God how He should run the world."

At the memorably contentious Solvay Conference in Brussels in 1927, Einstein and Bohr debated the implications of quantum theory for days on end before an audience of the world's greatest physicists. As a wearied Heisenberg recalled, "In the morning, Einstein presents a new challenge designed to prove the inadequacy of quantum theory. And every evening, after a day of hard thought, Bohr would have found some flaw in Einstein's reasoning." Still, Einstein persisted, creating difficult mind experiments that seemed to disprove quantum theory. Each day, with increasing difficulty, Bohr refuted his friend once again.

It is impossible to overestimate how much this controversy meant to scientists and others who trusted in the certainties of a Newtonian universe with its commonsensical laws of mechanics, electromagnetism, and thermodynamics. To some, the collision of the world of quantum mechanics, with its seemingly irrational suppositions, against the traditional worldview of classical physics was insupportable.

Contemporary physicist Michio Kaku has considered the effect of quantum mechanics upon religious thought: "According to traditional Judaeo-Christian thought, we have something called determinism. . . . God is omniscient. The quantum principle upsets the entire apple cart. If God cannot know where the electron is, he cannot know what a human being will do or feel in the future. Free will is reintroduced into our world. No book in heaven knows all of the future."

But quantum theory won out. Since the 1930s virtually no physicist doubts that this inexplicable world exists, despite Einstein's continuing objections, which seem more metaphysical than scientific.

Of course the strange behavior of the subatomic world does not change how humans deal with common objects. On a day-to-day basis Newton's laws efficiently describe how things work. Hurl a baseball at a window, and the pane breaks, even though, according to quantum mechanics, there's a tiny, tiny possibility that the ball could pass whole through the unbroken glass. But we cannot act upon such implications of quantum theory without ludicrous or even fatal consequences.

War Intervenes

The main reason for popular distrust of the new theories was the commonsense objection, but darker motivations rose to the surface in Western Europe in the 1930s. In Germany the Committee of German Scientists for the Preservation of Pure Scholarship publicly took the scholarly position that the inconvenient theory of relativity had produced a "Jewish physics"—in other words, a perversion of the authoritative "Aryan physics" of the late nineteenth century. Einstein was denigrated as a tasteless self-promoter, shouted down at lectures, and threatened several times with assassination.

Pictured on a mutual friend's veranda around 1927, Bohr and Einstein apparently sustained a respectful, affectionate relationship even though each struggled mightily to disprove the deeply held scientific convictions of the other. Bohr continued to refine his model of the atom by exploring the implications of his conceptual leap into the quantum world. Einstein, personally certain that the laws of the universe must be fixed and knowable, was never able to accept the apparent irrationality of that world.

(Oddly enough, this supposed far-out theorist had a part in inventing the gyrocompass, a very practical direction finder used by every navy at sea in the 1930s except the American and the British.) In a world of Newtonian consequences the most important theorist of twentieth-century physics decided to sail to safety in the United States. He never again saw the country of his birth. Settling in Princeton, New Jersey, he worked alone for the next three decades, almost universally revered but ever less relevant to developments in mainstream physics.

Under Adolf Hitler, public suspicion and resentment of Jews were nourished with the day's most sophisticated communications technology. In *The Eternal Jew,* one of many high-profile anti-Semitic propaganda films produced with generous budgets by the Nazi government, the narrator feigns a kind of evenhandedness: "The civilized Jews that we know in Germany give us an incomplete picture of their racial character. . . . The 'relativity' Jew, Einstein, hid his hatred of Germans behind his obscure pseudo-science."

Meanwhile the "civilized Jews," who represented a quarter of all of Germany's professional physicists, including eleven who had earned or were to earn Nobel Prizes, were forcibly retired from the civil service—that is, pushed into unpaid unemployment. Many fled westward. Other scientists enthusiastically joined the attack against four-dimensional space-time as "obscure pseudo-science."

The American Way of Physics

One surprise for the immigrant physicists was the American ease with machines as a tool for research. Ernest Lawrence, the hard-charging inventor of the cyclotron, virtually personified this American approach to physics. Born to modest circumstances in a small town in the Midwest in 1901, Lawrence had never had the luxury of ignoring the practical side of life. He could tear down and rebuild the family radio set and Model T. He put himself through college, where he majored in physics, by selling kitchenware door to door through farm country.

The concept of the cyclotron hit Lawrence in a flash in 1929, when he was teaching at the University of California. Trying to stay awake during a boring faculty meeting, he idly leafed through foreign-language scientific journals.

As he knew well, physicists of the day were investigating the basic structure of the atom by using a particle accelerator, a device that speeds up the movement of subatomic particles and smashes them against various targets. In this way, physicists could probe the inside of atoms and nuclei. The idea was to hit the nucleus with another nucleus, then examine the wreckage.

The first attempts at atom smashing drew upon high-voltage sources, including lightning. French physicists suspended an insulated cable between two peaks in the Alps and tried to divert the lightning's energy to a particle-accelerating tube that held an ion, an atom with a positive or negative charge. Thus accelerated by nature, it should easily smash the nucleus of a target atom. Unfortunately the tube could not contain the fifteen-million-volt lightning strike. One member of the team was electrocuted.

Other researchers tried more controllable sources of power, but the basic problem remained the same: It seemed impossible to tame and apply precisely the high voltages necessary to fire a nucleus with sufficient speed.

But the dozing Lawrence suddenly jolted awake in his chair. Inspired by an illustration not directly related to the issue, he suddenly realized that a circular particle accelerator, or cyclotron, can produce the required speeds by whirling particles around faster and faster. The spiraling path of the accelerating particle can be several times longer than the straight path of the linear accelerator.

Within this four-inch copper-encased cyclotron, one of Ernest Lawrence's earliest models, hydrogen protons are separated out inside a round glass box by an electric spark, then accelerated around the inner edges. The electromagnetic pole at bottom right and the next one clockwise to the left alternate their charges to keep the protons moving.

Lawrence had no doubts. "I'm gonna be famous!" he crowed to a colleague's wife as he dashed across campus.

He took the hands-on, can-do approach despite the modest means at his disposal when he began work in 1931. Parts for the landmark invention were scavenged from a radio, bought secondhand or cobbled together on-site. His work team's method of determining how much electricity was loose inside the jerry-built accelerator was crude. A nail was taped to a stick and warily brought close to the infant cyclotron until a spark leaped up to the metal. If the voltage was high enough, the spark might hit the experimenter's hand; if it was very high, he wound up in the emergency room of the nearest hospital.

Lawrence pressed forward relentlessly. When he finally collapsed into bed at night, he routinely turned on his bedside radio, which fortuitously picked up the low-

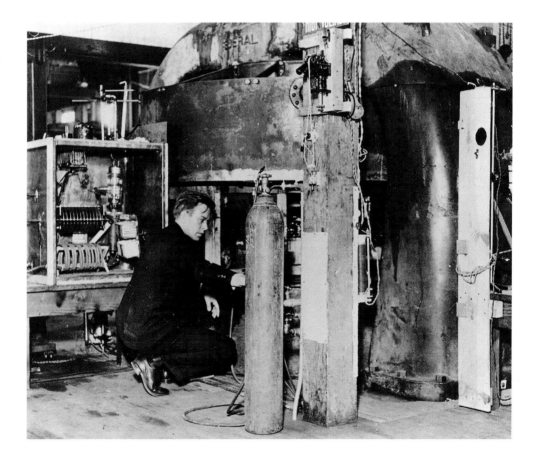

frequency radio signal generated by the cyclotron. If the air fell silent at that frequency, he was instantly on the telephone, furiously demanding to know why his invention was not humming along. He was obsessed with producing a million volts of energy to speed his particles to target.

His invention was simple but brilliantly conceived. First, air was removed from the circular cyclotron; then positively charged protons were introduced in a stream at the center. At opposite sides of the device were electrodes activated with alternating current. One electrode attracted the positive protons with a negative charge, then reversed to positive just as they arrived, repelling them. Then the protons were drawn to the other electrode, which had a minus charge. Once again the charge changed just as they arrived. The protons could simply race back and forth in a straight line in this fashion, but a magnet in the center forced them outward in a circular path. Their speed increased as they spun around the cyclotron. At the largest possible orbit they were flung out toward the target atom.

Lawrence's first machine worked, but he concentrated on building ever-larger cyclotrons. A tireless fund raiser for his projects throughout the 1930s, he became world-famous, just as he had predicted. Sometimes the celebrity veered toward notoriety; there were stories about his "death ray." But a *New York Times* editorial celebrated how his machine was a source of excitement and hope during the gloom of the Great Depression: Transmutation and the release of atomic energy "are no longer mere romantic possibilities. But it is the secret of matter that is of vital importance—the secret held by every star and stone. Fathom that, and the cosmos becomes an open book. Truths may be unveiled that we have been seeking ever since we started thinking about the Universe."

Lawrence won the Nobel Prize, but he soldiered on. It was now his dream to build a machine to produce energy levels of one hundred million volts.

"The Biggest Physics Project"

Meanwhile Einstein was followed to America by a flood of refugee scientists from Hitler's Germany and Mussolini's Italy: Enrico Fermi, Edward Teller, Hans Bethe,

The first cyclotron too large to fit on a table may look scatter-built by today's standards, but Lawrence was a meticulous, demanding inventor who used whatever materials could be converted to serve his inspired idea. Even the most powerful particle accelerators of today are the products of his essential breakthrough: the use of a magnetic field to accelerate and confine the motion of subatomic particles.

Eugene Wigner, and dozens of others. Their combined brainpower was to light up a growing community of U.S. theorists and experimenters. But the worlds of physics, politics, and military necessity were coming together to deal with yet another unanticipated discovery: In Germany in 1938 the nucleus of an uranium atom had been split, releasing a flood of energy and opening up the possibility of a chain reaction of atomic energy. A few short months later Niels Bohr secretly brought the news of nuclear fission to America, where the physics community immediately realized the horrific implications. Back in 1918 Rutherford had been the first to "split the atom," disrupting the nuclei of some nitrogen atoms, but this new event was no elegant demonstration in the lab.

When theorizing about the implications of relativity, Einstein had produced history's most famous equation, $E=mc^2$, in which E stands for energy, m for mass, and c for the speed of light. It suggested that a tiny amount of matter contains a staggering amount of potential energy. In fact the amount of energy will equal the amount of mass times the speed of light squared. The discovery of nuclear fission presented the possibility of creating a huge blast of destructive force, a leap of knowledge that could end all thinking in a flash. After being chased out of the German laboratory in which fission was first discovered, the Jewish physicist Lise Meitner worked with her nephew Otto Frisch to understand just what occurred theoretically in this extraordinary chemical reaction. Together they realized that fission was caused by the splitting of a nucleus of a specific atom, uranium 235. Most nuclear reactions—including the furnace of the sun, which stays alight by converting some forty-four million kilograms of mass into energy each second—are not very efficient, but nuclear fission could produce energy from elementary particles with devastating efficiency.

Scientists, military leaders, and statesmen outside Nazi Germany feared that Hitler's sophisticated, liberally subsidized physicists—heirs to the stunning theoretical tradition of the refugees now in America and Great Britain—might be able to design and build a doomsday weapon by creating a chain reaction of nuclear fission that could

After being ferried over dusty desert roads from the Los Alamos, New Mexico, experimental station in a Plymouth sedan, the plutonium that was the active material in the first explosion of an atomic bomb is gently carried into staff living quarters at the McDonald ranch house. Los Alamos was the heart of the Manhattan project, which employed forty-three thousand people at thirty-seven sites scattered throughout nineteen states and Canada.

release tremendous destructive power: a shock wave, terrific heat, and intense gamma and neutron radiation.

The United States hastily enlisted three quarters of the nation's most eminent physicists in the top-secret Manhattan Project. "Almost overnight," someone noted at the time, "physicists were promoted from semi-obscurity to membership in that select group of rarities which included rubber, sugar, and coffee." Whether originally from Europe or America, they become commodities in the U.S. defense effort.

The Manhattan Project was by far, in the words of a participant, Philip Morrison, "the biggest physics project of all time, and the future of the world depended on our success."

The project's leading scientists operated under the direction of J. Robert Oppenheimer at a special facility in Los Alamos, New Mexico. "It was unbelievably exciting," Morrison remembers. "One could see as many as eight Nobel laureates dining at once at the Fuller Lodge on the mesa." The average age of these ingenious intellectuals who profoundly altered humankind's attitude toward its capacity for survival was twenty-seven years. Most saw themselves as soldiers doing their duty in a war effort.

The government investment was huge, more than two billion 1940s dollars for the bomb project and a similar amount for improvements in radar. Only two years after the news about the chance discovery of nuclear fission had reached America, some of the largest factories ever built were devoted to making an atomic weapon. Harried researchers at Oak Ridge, Tennessee, and Hanford, Washington, worked to produce enough of the isotopes uranium 235 and plutonium 239, the most practical fissionable materials for the world's first explosion of a nuclear device.

Under intense strain, exacerbated by the fear of German superiority, and by the desire to solve tantalizing problems of physics and design, the young men at Los Alamos created a strong, even exuberant community. Edward Teller might drive his neighbors nuts by pounding the piano keys at two in the morning, Richard Feynman might tease the antsy military by picking the locks of combination safes that held classified documents, but they all were consumed by the technical challenges they faced. Essentially they were acting as engineers, but in a field without textbooks. Everything had to be invented on the fly, as they worked sixty- or seventy-hour weeks.

This was dangerous work too, as the young scientists experimented with fissionable materials whose properties were not yet intimately understood. Roger Warner, one of the engineers, recalls today how he and others put fears of explosions in perspective: "You can't run very far in a millisecond anyhow."

On July 16, 1945, a plutonium device was successfully exploded in the New Mexico desert. Three weeks later two so-called atomic bombs were dropped on Japan. Not since then has a nuclear device been exploded in war, but the shadowy image of the cataclysmically destructive "Bomb" began looming over international diplomacy and the nightmares of tens of millions of average citizens for the rest of the century.

Scientific researchers, looking for the nature of things, had accidentally discovered a phenomenon that technology could direct to destroy everything we held dear. Had we learned too much, probed too deeply, or were we just learning how to use knowledge well? The interplay between science and technology seemed to many just about equally exhilarating and disturbing.

Objects Strange and Wonderful

Ironically, just as the existence of life on earth seemed more tenuous than ever, there appeared hints of previously unknown forms of matter, or even life, thanks to an expanding array of methods of "looking" ever farther into uncharted space. During the war radar operators were annoyed by strange interference from outer space, but they had more immediate matters to deal with. Discovered in 1931, radio astronomy developed in peacetime, along with X-ray and ultraviolet astronomy, as a new way of investigating space and time. Although celestial objects radiate all the wavelengths of the entire electromagnetic spectrum, only radio and optical waves travel into human ken. The radio astronomer does not look through an eyepiece, of course, but waits for a computer to process distant radio waves and plot them on a graph. An antenna picks up the signals and converts them to electrical output, which is then amplified for storage on magnetic tape or for real-time display. The signals are almost undiscernibly weak after their voyages across light-years of curved space; according to one estimate, the total amassed by all radio astronomers to date would not provide enough energy to illuminate a flashlight for even a millionth of a second.

Back in Newton's bailiwick of Cambridge, astronomers in the 1960s set up a virtual forest of two thousand huge radio antennas on a soccer field and aimed them upward, scanning the entire sky above every four days as the earth rotated. Traditional astronomers needled radio astronomers for being interloping technicians who did not even know the locations of individual stars, but the new researchers were interested only in galaxies; that was where provocative radio signals seemed to emanate.

But what Cambridge research student Jocelyn Bell eventually heard from the heavens in 1967 was as unforeseen as it was inexplicable: a repeated radio signal as regular as a metronome. Was this the stellar imprint of an alien consciousness? Nature, in our world, offers no such unvarying, methodical pulsing.

Bell first noticed the bizarre object as a quarter-inch spike on the four-hundred-foot-long chart of the complete sky that the radio telescope took four days to reel out. In later charts she returned again and again to this "funny, scruffy, messy, unclassifiable

signal from the same bit of the sky." The signal was a string of pulses at regular intervals of one and one-third seconds. Logic dictated that the object had to be small, because it was fast, but in an apparent contradiction it was also large because the radio wave was so strong. In Bell's words, "Stars are great big things, like elephants, cumbersome, lumbering beasts, but this thing, whatever it was, was behaving like a flea, or some little jumpy insect. It was very nimble."

After all possible human-made signals were ruled out, there seemed to be no reasonable explanation. Indeed Bell and her colleagues met more than once to discuss the proper way to make an announcement: "Little green men," as they wryly put it, or some sort of extragalactic neighbors were sending us a message.

Such speculations later attracted the attention of the popular press, but the levelheaded Bell was relieved when she found a more rational, but no less wondrous, explanation.

A second type of signal was discovered, pulsing at the unvarying rate of one and one-quarter seconds on the opposite side of the sky from the first signal. The likelihood of two such alien transmitters was very slim. Then a clue came from physicists and their discovery of what happens to atoms subjected to incredibly intense pressure: They collapse into very dense matter. The source of the metronomic signals turned out to be a star that had crumpled from its gigantic mature form into a chunk about the size of an asteroid. When a giant star runs out of nuclear fuel, it first implodes, then explodes into a spectacular supernova. In the case of pulsars, the residue includes a neutron star, usually no more than six miles in radius but with more total mass than our sun. As the name neutron star might suggest, all of its electrons are crammed smack into its protons, thus forming neutrons with absolutely no open space in between. As this odd compact sphere rotates, it emits a powerful magnetic field in a narrow beam that sweeps across the galaxies, pinging regularly on earth's radio telescopes. Some pulsars have been clocked spinning as fast as one thousand times a second. By discovering these previously

Aimed toward the farthest reaches of space and time, the twenty-seven highly sensitive dish-shaped antennas of the National Radio Astronomy Observatory's Very Large Array (VLA) in Socorro, New Mexico, can be focused by moving them on rails. If close together, they scan a wide part of the sky; when spread out for twenty miles, they concentrate on details in a small area. The inset photo, a computer-processed image, is the most detailed radio image yet of the gas in the spiraling arms of a galaxy in Ursa Major known as M81.

unimagined stars, the new field of radio astronomy proved itself capable of revealing cosmic wonders where not a thing could be detected by human eye or optical telescope.

It might seem at first glance a haphazard relationship, physicists and astronomers occasionally joining together to discover yet another uncanny object in the skies, but the list of celestial marvels accelerated in the second half of our century. For example, there was the theoretically seductive event known as a black hole, a concept that has launched jittery scenes in novels and movies like the following climactic moment in Hollywood's *The Black Hole:*

Ernest Borgnine
My God, right out of Dante's Inferno!

Anthony Perkins
Yes. The most destructive force in the universe.

Richard Forster
Nothing can escape it, not even light.

In 1967, when graduate student Jocelyn Bell first noticed that the radio telescope she was using at Cambridge University had detected a regular pulsing signal deep in space, she wondered playfully if she had stumbled upon proof of intelligent life on another planet. In fact, she had discovered the first of several hundred pulsars known today. These incredibly dense collapsed stars send out narrow bands of energy that register as pulses on the recording equipment of radio telescopes.

Yvette Mimieux

I had a professor who predicted that eventually black holes would devour the entire universe.

Anthony Perkins

Why not? When you can see giant suns sucked in and disappear without a trace?

Richard Forster

It's a monster, all right.

Robot

A rip in the very fabric of space and time.

In June 1997, the orbiting Hubble Space Telescope photographed the near vicinity of a black hole for the first time, using an imaging spectrograph. In this astonishing false-color image from a series of photos, the twin cones of gas emission caused by the release of energy from a supermassive black hole are streaming out at velocities of hundreds of miles an hour. The white horizontal line is the radiation beam from the black hole seen sideways to the line of sight from the earth.

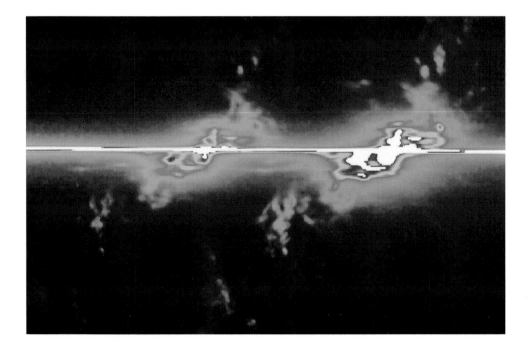

This intentionally simplistic dialogue is not so far from the unnervingly complex truth. Einstein's general theory of relativity mandates the existence of black holes: When a giant star implodes, it might become too weak to flare up again into a supernova. Gravitational forces cause it to collapse upon itself, perhaps shrinking into a pinpoint of infinite density or into a gigantic, invisible object without a surface. Even experts find it difficult to visualize black holes, but their effects are clear. Light cannot escape from the powerful object. Space curves toward it, sealing off the star's matter, as titanic clouds of hot gases and living stars swirl violently around it. In some cases, jets of high-energy particles are forced out from the black hole in opposite directions.

To date the evidence for actual black holes is indirect, but at least six may have been found by analyzing the effects of their massive gravitational fields. Invisible sources of X rays that have been located near double-star systems, for example, are perhaps produced when the gases from a visible star are sucked with tremendous force into its dark twin and then heated and compressed enough to emit X rays.

There are other odd ducks in the sky. As diverse signals speed back and forth over vast distances, then are caught and fixed and interpreted on plates and slides and paper, celestial reality outstrips previous flights of human imagination. Far, far off, first discerned by infrared wave techniques in 1963, are the most luminous and most distant objects of the known universe, the faint blue quasars, or quasi-stellar objects. They are apparently racing away from us at something like 80 percent of the speed of light, and they are already about eight billions of light-years in the distance. To be faintly visible from this perspective, they must be as bright as one hundred galaxies put together. No one yet knows what they are.

Even in the relatively small portion of sky each of us can see for ourselves on the clearest night in the Great American Desert, we are unconsciously traveling in time with every glance. The light of a star in the cowl of heaven may shine there billions of years after it left its source, but the star itself may long be dead. We gaze across a sea of stars that exist or have existed light-years and eternities apart from one another.

Big Bang Versus Steady State

Ever since Hubble's discovery in the 1920s that galaxies are moving away from one another at tremendous speeds, astronomers have speculated about first things. If we imagined the process in reverse, many physicists reasoned, we would find ourselves back at the beginning of time. For these billions of star systems to be racing outward so many billions of years afterward, there must have been an initial cosmic explosion. This conjecture became familiar to scientist and layperson alike as the big bang theory.

Evidence from more than one source seemed to support the idea, which posited the birth of the universe somewhere between ten and fifteen billion years ago. There are some minor problems with the theory, and one major one: What existed *before* the natal explosion? In the 1950s, the English astronomer Fred Hoyle countered the big bang enthusiasts with the hypothesis of a steady state universe.

He found big bang "just not dignified, or elegant . . . rather like a party girl jumping out of a birthday cake." What he preferred, and what his theory implied, is a universe that has existed for all time, "with neither a beginning nor an end."

As Hoyle's gently wry explanation suggests, the choice between the two theories was often consciously a choice made for psychological or aesthetic reasons. For more than a decade, no conclusive evidence had been found to demolish either hypothesis; indeed the scientific community in the 1950s remained pretty evenly divided between them.

Ironically, the probable definite answer appeared almost simultaneously in two different places in 1964: in one case as a sudden inspiration and in the other as an irritating technical problem. At Princeton astrophysicist David Wilkinson was working with colleagues in the lab one afternoon when group head Bob Dicke rushed in, wiped the blackboard clean, and began furiously scrawling. He had figured out, and was proving, that a big bang would produce certain radio waves in the microwave range, where wavelengths range from 0.04 to 12 inches, and the echo should be detectable these billions of years later because the birth of creation was so intensely hot.

This beautifully simple idea did not immediately excite Wilkinson, but he and colleague Peter Rowe at Princeton worked for a year to put together a radiometer capable of tracking down creation's voiceprint. Because Dicke was fervently thrifty, the instrument was made from cheap parts scrounged from World War II surplus shops in Philadelphia.

The day they prepared to take their radiometer up to the roof to begin their experiments, however, a weary call came from the Bell Laboratories thirty miles away, where radio astronomers Robert Wilson and Arno Penzias were near the end of their tether. For a year the Bell researchers had been struggling to get a clean signal from their satellite communications antenna. Its constant hiss interfered with their attempts to observe the universe. They could have subtracted the hiss from their data, as others commonly did, but they refused to. Even scrubbing away the droppings produced by a pair of nesting pigeons did no good.

Dicke instantly recognized that this "interference" was an observation of cosmological significance. By accident, as Dicke and his Princeton counterparts had been gearing up to prove big bang, Wilson and Penzias at Bell had overheard the ten- to fifteen-billion-year-old signal that radiates outward from the explosion that created everything. Dicke told his team: "Well, boys, we've been scooped."

Ironically, Wilson and Penzias had supported the steady state theory. They were undoubtedly happy to accept the Nobel Prize, however, earned because their bothersome background noise was considered proof that the universe had begun with a big bang. The cosmic microwave background they heard is the afterglow of that primal explosion. It is radiation that has not interacted with matter in ten billion years. It is, in other words, a picture of the universe as it actually existed about three hundred thousand years after the big bang.

Yet questions remain. Will the entire universe expand until it becomes exhausted, then collapse back upon itself like a dying star? This would be a so-called closed uni-

verse. By contrast, an open universe would go on expanding forever, another credible possibility. Or has the universe attained an equilibrium between competing forces so that its rate of expansion nears zero? That would be a flat universe, the most popular model today. But if it is finite, even as it expands, what else is there?

Indeed, the big bang theory is pockmarked with enough nagging inconsistencies to keep its advocates and a few detractors busy thinking and experimenting for the foreseeable future of physics and astronomy, which, as we see again and again, has not always been at all foreseeable throughout this extraordinary century.

In 1979, a postdoctoral fellow in astrophysics at MIT, thirty-two-year-old Alan Guth, produced the most important new idea in cosmology since Hubble's original concept: the idea of inflation. Guth saw that an extremely rapid expansion—an intense period during the very first second of the big bang, completed within that ancient second—followed by today's slower rate of expansion could explain how the universe could now be flat.

Guth's clever insight was refined and incorporated into the current standard model of the big bang: In the first trillion trillionth of a second of time, the grapefruit-sized universe, some one hundred billion degrees hot, contained matter and antimatter, electrons, positrons, photons, and the mysterious massless neutrinos—all violently bumping against each other and collapsing and destroying and re-forming. Physicists now believe that these earliest of interactions determined the large-scale structure and ultimate fate of the entire universe.

The Fundamental Laws of Everything

The tiny particles at the beginning of time still form and define all things large and small. Since the entire universe visible to us was once, however briefly, on a subatomic scale at the moment of the big bang, it can be understood only on the level of subatomic or particle physics. In the latter half of our century, as astronomers and astrophysicists studied time and space on the grandest scale, physicists concerned with the fundamental forces and elementary substances of matter drove ever deeper into the infinitesimal unknown. Theirs is the world of microphysics.

The researchers using this horn antenna to track one of the first human-made satellites, Echo I, *became increasingly annoyed by a persistent background static in the device, which was part of the Bell Laboratories at Crawford Hill, New Jersey. No matter where the antenna was aimed, the noise continued. What Arno Penzias and Robert Wilson eventually realized was that they were hearing the background radiation from the birth of the universe in the so-called big bang.*

Fueled by the political and military anxieties of the Cold War, as well as by "friendly" competition between the United States and Europe, the research of micro-physicists progressed rapidly in the 1960s. Ernest Lawrence's home-built cyclotron cost about $350 to build, but the accelerators of high-energy particle physics cost billions. Instead of a few scruffy researchers working together, a problem required the work of hundreds of researchers, and an experiment might last for decades.

Whatever governments might want, high-energy particle physicists continued to dream of putting together a unified theory—in effect, an elegant single explanation of the complex behaviors of all matter and all forces that exist in the universe.

But the particle accelerating machines eternally revealed new kinds of elementary particles. Smash an atom, tally the wreckage, and yet another discovery is made. In an additional legacy from Einstein's famous equation, $E=mc^2$, the energy of the collision in particle accelerators was creating new matter, revealing new atomic particles. In improved detectors, known as bubble chambers, the particles leave vapor trails that were automatically photographed when a sensor spotted them. These oper-

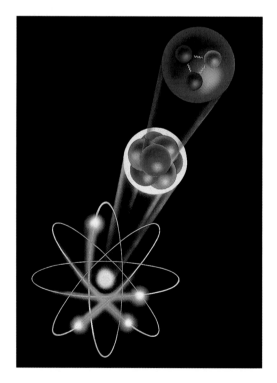

An exploded view of an atom reveals what physicists have learned by the end of the twentieth century about the inner structure of subatomic particles that were once thought to be fundamental units of matter. In the center, the nucleus of an atom of beryllium is illustrated according to today's theories: The four electrically positive protons colored red in the nucleus are balanced against the four orbiting negative electrons, while three of the atom's five neutrons are visible in blue. Protons and neutrons are both constructed of three quarks. At the top, a proton is shown with its single "down" quark and two "up" quarks. A neutron has one "up" quark and two "down" quarks.

ations were incredibly swift, since most particles live less than a millionth of a second.

Back in the 1930s the few known particles included the small squad familiar to many of us from required high school courses—primarily the protons, neutrons, and electrons. By the time of the Nixon-Kennedy debates in 1960, more than 50 particles had been counted. There are finally more than 150 toted by the 1990s, including such things as muons and taus. Naturally this glut called into question the very notion of "elementary."

The scientist who made sense of it all was Murray Gell-Mann, who by age twenty-six spoke eight languages, apparently well, and was a full professor of physics at the California Institute of Technology, or Caltech. As cheeky as he was intellectually nimble, Gell-Mann was often called the smartest man in the world.

Taking a cue from classical biology, he began classifying subatomic particles into families in the late 1950s, the better to understand their characteristics. His lists predicted the existence of an unknown particle, the omega-minus. At Brookhaven National Laboratory, after years of tediously reviewing photographs taken in accelerators, the young physicist Nicholas Samios finally discovered the omega-minus in 1964 on experimental photograph No. 97,025. Gell-Mann's classifications were on target.

Next, he proposed that most forms of matter must be composed from combinations of fundamental, if unknown, particles, which he called quarks. The nonsense name is playfully based upon a line in James Joyce's "difficult" novel *Finnegans Wake:* "three quarks for Muster Mark." Gell-Mann was similarly antic in suggesting six different "flavors" for his supposed quarks: up and down, charm and strange, and bottom and top. The great majority of subatomic particles, he believed, must be formed from some combination of these quarks. For example, it takes one down quark and two up quarks to make a proton.

The basic idea was so attractive that other scientists began designing experiments to prove the thesis. One of Gell-Mann's friends ground up oysters since they tend to ingest virtually every chemical element in the seas, while other researchers examined cosmic rays at very high energies.

For ten years the results were disappointing, and quarks were roundly dismissed in some quarters, not always without malice, as mere figments of Gell-Mann's imagination. For one thing, they would have to have a fractional charge, unlike the electron's or proton's charge of unit one. Traditional scientists did not believe that fractional charges were possible, much less suggested by available evidence.

But new data from the Stanford Linear Accelerator, collected in the late 1960s, seemed to reveal the presence of hard scattering centers within both protons and neutrons. Just as Rutherford had been stunned by the recoil of his alpha particles off atomic nuclei five decades earlier, many physicists were convinced by these results that quarks really do lie at the center of most types of subatomic particles.

The brilliant Gell-Mann had been proved right. The scientific community agreed that the smallest elementary particle is the quark. Five were found by the end of the 1980s, but the heaviest, the top quark, remained elusive for years. Physicists did not catch their first glimpse of this sixth quark until late in 1994.

The Standard Model

Now attention focused on another basic issue: How can the four forces that keep the universe and its constituent parts spinning along be conceptually unified?

Two of the forces, gravitation and electromagnetism, are experienced in everyday life. The other two forces, known simply as the strong force and the weak force, affect only the particles inside each atomic nucleus.

For anyone attempting the conceptual leap that would produce a unified theory involving all four forces, it is surely daunting that Einstein himself spent the last three decades of his life on the problem but failed.

Light began breaking in 1967, when the electroweak theory was proposed independently by Steven Weinberg and Abdus Salam. (Earlier, many essential elements of the model had been pieced together by Sheldon Glashow.) Put succinctly, they assumed that electromagnetism and the weak force in atomic nuclei were the same force in the first moments of the big bang. Then, as the universe cooled, these forces were separated. For this to work, the existence of two then-unknown particles had to be assumed. They were called the W and the Z.

At the European Center for Nuclear Research (CERN) accelerator in Geneva, Switzerland, the legendary, exasperatingly energetic physicist Carlo Rubbia was determined to find those missing particles, which he called "the little beasts." Twice before,

Set in operation in 1965, the two-mile-long accelerator of the Stanford Linear Accelerator Center in the San Francisco Bay Area of California continues the tradition of research initiated by Ernest Lawrence in his four-inch cyclotron. From the lower right-hand corner, electrons are accelerated to high speeds for atom smashing and other experiments at the other end. Because low-mass electrons lose energy in a cyclotron, however, the SLAC was built to beam them in a straight line.

his remarkable work had only narrowly missed winning the Nobel. In the early 1980s, therefore, he was now highly motivated. His tool was CERN's European bubble chamber, twelve feet tall, which held 1,350 cubic feet of liquid hydrogen through which protons could be accelerated to very high speeds. The collisions within could be recorded on very fast film by four separate cameras.

Rubbia was a gifted, persuasive politician who skillfully played upon European resentments; for two decades all of the Nobelists in high-energy particle physics had been Americans. To move the research ahead, each new accelerator had to be more sophisticated—and more expensive—than its predecessors. Rubbia had to be a brilliant theorist, inspired leader, and relentless politician. He earned the name Alitalia Professor because

The elusive W and Z particles were found with the aid of this detector in the enormous proton-antiproton collider at CERN, the European Organization for Nuclear Research. A team led by Carlo Rubbia decided to seek out these particles, crucial to the electroweak theory, in head-on collisons of protons and antiprotons, an approach that some skeptics expected to produce little more than a mound of subatomic rubble.

he flew the airline so frequently in search of funding. Friends joke that when he stood still, his average altitude had to be two thousand feet and his average velocity 40 mph. The result: one hundred million dollars for CERN to build a radical new accelerator.

The money allowed Rubbia's team to convert one of the center's existing accelerators into a machine for smashing protons and antiprotons together rather than, as in conventional experiments, shooting protons at atoms. To more conventional physicists, this was a doomed, if not deranged, exercise, akin to slamming two garbage cans into each other at high velocity and expecting to make sense of the resulting rubble. To snare either the W or Z, Rubbia had about a billionth of a trillionth of a second before they would blow apart in a shower of subatomic debris. He added a twenty-million-dollar detector the size of a mansion to photograph virtually every vagrant particle. This experiment may have been the most complicated ever in particle physics.

"The world's leading accelerator physicists said it wouldn't work," Rubbia recalls. "These remarks were coming from very, very good people. Nobel Prize winners. I was scared stiff they were right."

Perhaps. But those who knew something about the man and his career were not too surprised when, on January 20, 1983, he and his gigantic team of researchers found the W particle. The next day physicists working with another CERN accelerator duplicated their results. The CERN staff found the Z within months. Rubbia pocketed his Nobel, and physicists finally had their so-called standard model, which pulled together quarks and the four forces in an outline for how the universe works.

Unfortunately, this model, essentially a page listing the various forces and particles that apparently construct the entire universe, is considered too disorderly, cluttered,

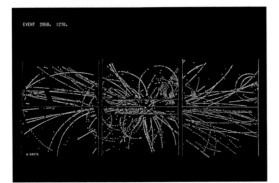

This color-treated photo shows a computer reconstruction of the tracks left by particles in a head-on collision in the CERN proton-antiproton collider in late 1982. The arrowed line indicates a high-energy electron that emerged in conjunction with so-called missing energy, which would be emitted by invisible neutrinos. Rubbia and his associates realized that this subatomic debris resulted from the decay of a W particle.

and complex by most scientists, who live by the classical ideal that the truest explanation is the most elegant.

"We are drowning in quarks," Michio Kaku has complained. "The standard model, with all its conceptual ugliness, is only a transition, only the first step toward the final theory."

That theory may not be so far off, thanks to the continuing work at the world's most powerful particle accelerator at Fermilab (Fermi National Accelerator Laboratory), which was founded in 1969 in Batavia, Illinois. Nearly a century after Rutherford's scattering experiments with alpha rays, scientists eager to discover how the world works can propel eight hundred billion protons at a time in a fierce stream around a four-mile-long circular machine at 99.99995 percent of the speed of light. As Einstein taught us, the speeding subatomic particles grow millions of times more massive. In a collision four hundred million times hotter than the furnace of our sun, a trillion of a trillionth of a second that re-creates an instant of the big bang, a part of this huge energy becomes matter, again as Einstein predicted. In that inconceivably brief moment, Fermilab is transformed into a kind of time machine, taking us back to the blast that formed space and time, and perhaps providing clues that will soon produce a final theory both accurate and graceful.

Because of such mammoth projects as Fermilab, physicists have long assumed that they would continue to work hand in hand with nurturing governments until every mystery of atomic structure was unequivocally resolved. The jewel of technological instruments in high-energy particle physics was to be the twelve-billion-dollar superconducting supercollider, a fifty-mile-long circular tunnel that would be the largest experimental tool ever built. The United States, fearful that Europe was taking the lead, decided to go with the project as soon as the Z particle was discovered. Two streams of protons could be made to race around the supercollider in opposite directions at near the speed of light, then collide in an explosion of a trillion electron volts.

Then the Cold War ended. In 1993, after spending two billion dollars preparing a site in the scrub desert near Waxahachie, Texas, the U.S. Congress canceled the project. Many physicists were instantly downsized out of their field; others continued to work at the aging accelerators still in operation.

Twenty-first Century Science

Relativity, general and special; quantum theory; quarks; and the four forces: By the last days of the twentieth century they seemed almost cozily familiar in the popular view of the universe. The movies had absorbed "lite" versions of them all, and casual references in the press or in conversation used or misused them; they had lost much of their power to disturb us with their insinuations of irrationality and the imponderable.

But some theoretical physicists will not let it rest there. They have come up with string theory. It may not be true, or it may lead at last to the discovery of the unity of all forces with all of the matter in the universe. It is so complex in its details that the majority of physicists lack the background for understanding it, and many therefore regard it with nervous hostility. Yet it just could be the ultimate theory, a concept possibly so far ahead of its time that it cannot be accommodated fully until the next century.

Some parts of the picture don't entirely resist description. Essentially, string theory suggests that the fundamental unit of matter and also of all the natural forces resembles a string. When open, with two ends, the string shape is the basic unit of a force; when closed in a loop, it can be the basic unit of matter, perhaps one hundred billion billion times smaller than a proton.

No other surmise about fundamental particles allows for the action of gravity. String theory does. Moreover, and most beautifully to the mind of a theorist, strings vibrate, thereby generating the characterizing properties of all known subatomic particles.

As string theorist Sylvester James Gates of the University of Maryland sums it up, "All particles of matter and energy are but different harmonies of strings."

Lovely, but not that simple. The theory does not actually envision strings as we know them in the visible world but instead as tiny, one-dimensional rips in the smooth fabric of space-time. They are infinitesimally thin, perhaps best described as "pure length." In addition, the theory requires that these "strings" have to exist in at least ten dimensions, when the fourth has taken decades to grasp. Human beings have no way of accessing dimensions five through ten. Such concepts can only be discussed in the language of pure mathematics.

"We've got that covered," Gates says. "Those other dimensions are curled up so tight you'll never have to deal with them." Nor are Gates and other compelling enthusiasts troubled by another little difficulty: Experimental confirmation of string theory might require a particle accelerator roughly the same size as our solar system.

But for Nobelist Sheldon Glashow, a prominent dissenter, the string theorists commit the cardinal sin for physicists: the separation of theory from experiment.

"String theory is in danger of drifting away into the ionospheric reaches of pure, abstract thought," he has complained. "Contemplation of strings may evolve to the point where it will be conducted at schools of divinity by future equivalents of medieval theologians. For the first time since the Dark Ages, we can see how our noble search may end, with faith replacing science once again."

The controversy rages on, but it is for now only a small part of the huge accumulation of information and speculation that has changed the way we have viewed the physical world in the last hundred years.

Whether string theory proves to be a dead end or a revelation that can be considered cosmic in its reach and consequences, it is clear that we probably know much less than we think. That has been one of the major lessons of our century. Science could almost completely describe the known universe in the 1890s, but that was the rub, for what was *known* has been disproved or shown to be inadequate overnight, and then again, and again, and again. Discovery in our day is exponential.

Fate of the Universe

Foolish as it may be to try to infer the future of discovery in physics and astronomy, there are certain hints. The latest generation of telescopes, larger than any ever built before at twenty feet in diameter, form a battery of machines producing photographs of unprecedented depth and clarity. Larger ones are planned, their capabilities multiplied several-fold by such new developments as adaptive optics, a laser-based technique for eliminating the visually disturbing effects of turbulent air. The Hubble telescope now orbiting the earth—the first device to scan the heavens without having to deal with the distortions of atmosphere—will be followed by orbiters of greater sophistication.

In search of more precise information from radio waves, huge concave "dishes" have been constructed to capture and reflect the signals to a dipole antenna set up in the center. A fixed reflector dish one thousand feet in diameter and eighteen acres in area probes deep into space and time from a depression at Arecibo, Puerto Rico. It has made such extraordinary discoveries as ice in the polar regions of the planet Mercury and the existence of planetary systems far from our own. Built in 1963, the radio telescope's sensitivity to remote radio-emitting objects was quadrupled by a $27 million, five-year-long upgrade completed in 1997. Researchers expect its enhanced capabilities to retrieve information about the actions of galaxies billions of years in the past and also detect asteroids before they strike the earth in the future. It automatically checks trillions of signals for evidence of extraterrestrial communication and will analyze one of astronomy's most recent surprises—the so-called microquasars in our galaxy that give off huge amounts of radio waves and giant spurts of matter.

In addition, the Keck Telescope and its associated observatories more than fifteen thousand feet high on the slopes of dormant Mauna Kea volcano in Hawaii are raking in new information. Four times wider across than the next largest telescope in the world, the Keck is some 393 inches in diameter and weighs three hundred tons. Like a giant and highly polished insect's eye, it is composed of thirty-six individually focused mirrors, each six feet across. The dome of the telescope is kept near freezing to prevent temperature variations that would distort the glass and steel parts. Photons captured by

The thousand-foot-wide radio telescope disk at the Arecibo Observatory in Puerto Rico, the world's largest, was recently upgraded to provide the most sensitive ear on the planet. It captures extremely weak but highly penetrative radio signals that are projected from the depths of space by any object warmer than minus 459.67 degrees Fahrenheit, or absolute zero.

the mirror in visible and infrared wavelengths of light are conveyed to different kinds of electronic detectors to produce and interpret images.

At telescopes around the world, researchers at the end of the century were photographing supernova explosions on a regular basis. Supernovas, which are as bright as a billion stars, can be seen halfway across the universe, but they flare up and dim out within a month or so, making rapid measurement necessary. By comparison with the oceans of stars in existence they are extremely rare—our own galaxy of one hundred billion stars averages one per century—but the various participating telescopes can together cover several thousand galaxies. By comparing photographs of a section of sky taken two weeks apart, the project can often find as many as ten of the giant stellar ob-

jects in a single night. The information is sent over the Internet to a facility in Cambridge, Massachusetts, then distributed worldwide to scientists engaged in trying to measure the geometry of the universe.

Why is this measurement so important? Because it turns out that the shape of the universe, its eventual fate, and the amount of its mass are related. Most of the matter in the universe is invisible, or so-called dark matter, but no one knows what it is. It is known to exist only because, using Newtonian physics, it is possible to estimate the weight of the universe. The result: The visible universe is a mere fraction of the total weight, for somewhere between 90 percent and 99 percent of the material of the universe that can be weighed is missing. Apparently, it is unlike any matter that we know or understand.

But whatever it is, its total weight can perhaps be measured most accurately by finding out how much the dark matter bends space, thus causing light to curve. That weight may be cosmically predictive. As the universe expands, the total amount of dark matter would have a determinative effect upon the process: It would either slow it down, bring it to a halt, or reverse it. In other words, the dark matter may determine whether the universe expands forever, reaches an equilibrium, or collapses back upon itself in the so-called big crunch.

In theory it should be possible to measure the degree to which space is curved by comparing the brightness of a supernova with its distance as indicated by the red shift in its spectrum. The ratio of brightness to distance, in other words, might reveal the fate of our universe.

In only one hundred years science has conceptually flown from the quaint dry watercourses of next-door Mars to stars visible ten billion years back in time, from solid objects falling always smartly to the ground to gaping black holes that suck in carpets of space and time, and from buzzing little atoms to strings in ten dimensions uniting matter and all the forces.

All this is only prologue.

Orbiting around the earth since April 1990, the Hubble Space Telescope, a 2.4-meter reflecting device, produces extraordinary images of deepest space, free of the atmospheric distortions, light pollution, and other types of interference that mar optical telescopy. Resolution is ten times greater on average than with ground-based telescopes, allowing scientists to determine the age and distance of previously unknown galaxies at the farthest edge of the universe.

Six airplanes flying with a dirigible were presented as vehicles of light entertainment at Southern California's Aviation Park in 1910, but world war would soon reveal heir darker potential.

BIGGER, BETTER, FASTER

We take for granted lives of convenience that would have been considered unimaginable luxury throughout human history until about 1900. In fact we often tend to focus on the negatives left in the wake of the twentieth century's technological odyssey: the cost, damage to the environment, loss of community feeling, materialistic excess.

We are sometimes glib about technological marvels, but it would be a great loss if we could not recapture the sense of wonder that amazed our great-grandparents as they experienced the technological flowering that began some hundred years ago. As they saw so clearly, it conquered many of the ancient factors that made human life, for most people, a continual round of hard work, inadequate diet, extremes of temperature, and deadly dullness.

What explains this explosion of technology? Perhaps the greatest single error most of us make in considering twentieth-century technology is the traditional assumption that the story of technological change continued to be a series of portraits of ingenious or lucky inventors. In America we are still the enthusiastic heirs of a lively tradition of individualism. Our history has been told again and again as the lives of saints and sinners.

But this is not the way the technology that has defined our century was developed. While there have been extraordinary innovators, they have typically succeeded as members of a team or as heirs to a developing technique or methodology, and they have changed our lives because their achievements have been promoted by visionary business leaders, encouraged by academic institutions, and often funded by agencies of government.

In other words, technology has not been the extraordinary invention alone; technological growth sprang from the systems and organizations that marry the new product to society's needs. Fundamental to progress there has evolved a kind of up-

ward spiral of linked development: first, the accelerating growth of new consumer goods created by technology, then mass production of the most successful goods through technology, then sales in vast numbers spurred by the technologically innovative new medium of advertising, which have in turn continually rekindled the demand for even more new technology. Above all, competition has shaped technological progress, as we'll see over and again in this chapter.

Taking Off

In 1903 several thriving Southern California newspapers, like their counterparts elsewhere, ran vague stories about some kind of "flying machine" taken briefly aloft by Wilbur and Orville Wright on a hill beside a North Carolina beach. The world scarcely noticed. For one thing, the inventors were secretive about the exact nature of their invention; for another, the stunt did not seem relevant to everyday life.

The Wrights, who owned a Dayton, Ohio, bicycle shop, had been puzzling over the challenges of manned flight since at least 1896, when Orville was twenty-five and Wilbur twenty-eight. Some four decades before, in 1853, the English inventor Sir George Cayley designed a piloted glider that sailed briefly in the air, logging history's first gliding flight. In the 1890s, the German aviation pioneer Otto Lilienthal learned a great deal about controlled flight in more than two thousand gliding experiments. But he was killed when an unexpected strong gust of wind caused his glider to stall in midair and tumble to earth. Meanwhile, various inventors had managed brief, unpromising hops with powered craft, but the chief obstacle defeated all: It seemed impossible to control the plane in the air. The Wrights solved the problem. They carefully observed the flight of a buzzard and realized that the bird retained balance in the air by twisting the tips of its wings. Their so-called wing warping method mimicked the actions of wing feathers through a system of interacting pulleys and lines. The pulleys were attached to the wings and controlled by the pilot with the lines. In this way the wing ends could be bent and the aircraft stabilized as it maneuvered in flight.

This device resulted in a remarkable degree of maneuverability. The brothers had

At the 1900 Grande Exposition Internationale in Paris, the industrializing world celebrated such triumphs of technology as the laborsaving electrical motor and the Eiffel Tower, at 984 feet the world's tallest structure, while motor-powered balloons proved French dominance of the air. Within only a few years, a flood of technological innovations would make this fair seem quaint.

recognized that a plane will turn gradually if it is banked, not unlike the method of turning one of their bicycles by leaning to one side. By warping one wingtip upward and the other downward at the same time, they produced a slight roll. At first their method was unstable. Raising a wing produces more lift, but it also produces air resistance, and the plane swerved out of control, even crashing. By adding a vertical rudder to produce a slight skid, however, the Wrights learned to bank and turn in perfect control. Their discovery—a breakthrough that for the first time made controlled, sustained flight possible—is, in modified form, basic to controlled flight today.

Designing their own propeller and a gas engine that would fit their invention, the brothers built a plane, the *Flyer,* that flew for twelve seconds above North Carolina's Outer Banks at Kitty Hawk on December 17, 1903. Each succeeding version of the *Flyer* flew longer as the Wrights continued their experiments.

Profit-minded businessmen in Southern California were not slow to see the po-

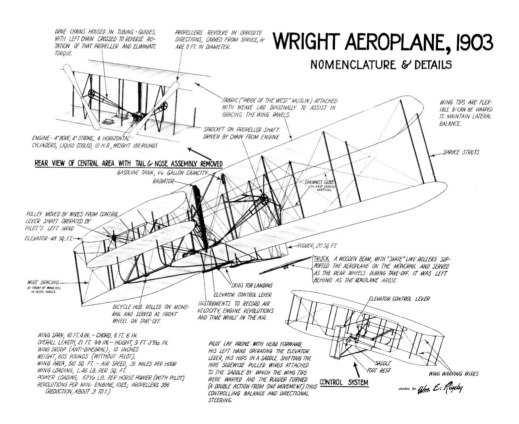

tential for a marriage of this dramatic achievement and the development of their desert kingdom. Just two years after the Wrights' first public demonstration of their plane in 1908, L.A. boosters sponsored the first major international air competition ever held in the United States. The optimistic organizers built grandstands for up to thirty thousand people and plastered Southern California communities with posters for this unprecedented event at the newly christened Aviation Field south of Los Angeles.

If there is such a thing as pure sports, this was not it: the substantial prize moneys were meant to attract competitors who had discoveries to show off and experiments to subsidize. From these showboating pilots to the huge aircraft manufacturers of the future, competition would always be the mother's milk of aviation progress. Flight attracted people who wanted to go higher and faster and farther.

The extraordinary sum of fifty thousand dollars was paid to the colorful French pilot/circus performer Louis Paulhan so that he would attend the event and demonstrate the capabilities of two important new airplanes: the Farman biplane, designed and built in Britain, and the French Blériot monoplane. Paulhan's well-turned-out wife and her French poodle added a touch of exoticism to middle-class Southern California, and a lawsuit provided a touch of scandal: The Wrights claimed that both planes infringed upon the copyright of their latest craft, the Wright *Flyer*.

Kevin Starr, California's official historian in the 1990s, has explained how this ingeniously commercial occasion had a special meaning for the development of technology: "The air meet was an extraordinarily sophisticated public relations gesture on the part of L.A. county. . . . It was as if [local boosters] saw through a glass darkly the whole impending 20th century drama of technology anticipated in those aircraft."

Not surprisingly, a crowd of curious onlookers filled the stands at Aviation Field. Many were able to attend only because the Pacific Electric trolley line ran special cars from downtown. Most had never seen an airplane in flight, and quite a few did not expect to do so on this occasion.

Glenn Curtiss, known as a roughneck motorcycle racer skilled at designing ma-

This drawing explains some of the features that made the 1903 Wright Flyer *the first craft capable of powered, sustained, and controlled flight. Building upon the achievements of many gifted inventors, Wilbur and Orville Wright recognized that the principles of riding a bicycle upright could be applied to the problem of steering a plane without causing it to swerve out of control.*

chines, swiftly proved them wrong. The first American to fly after the Wright brothers, he had already captured the world's airspeed record, an astonishing 46.5 miles per hour.

When his yellow-winged biplane was ponderously hauled out in front of the grandstand, the spectators fell silent. The self-confident Curtiss posed for photographers, then signaled a mechanic to spin the propeller until the engine caught with a roar. The crowd cheered. But it fell silent again in awe as the fragile biplane bumped along the ground, rose swiftly to a height of fifty feet, then described a circle of about five eighths of a mile in one minute and twenty-eight seconds.

When Curtiss landed in one piece and came safely to rest a hundred yards from his takeoff point, there were doubters no more, and the grandstand rocked with wild cheering. It was the kind of flabbergastingly incredible exhibition that had strangers shaking hands with each other, jumping up and down in excitement. For the first time in their lives they had just seen another human being wing up into the air swiftly and in control of flight, dramatically freed from the grasp of the earth.

Curtiss was not the sole advertisement for the future of humankind in the air. For eleven days the air show introduced aviators and aircraft, each performance a stunner by its very existence and many of them record-setting. But it was the flamboyant Paulhan who captured the prize for showmanship. With his wife he flew twenty-one miles across the developing countryside, breaking the world record for a flight with a passenger aboard. He soared higher than any other competitor, climbing to a breathtaking forty-one hundred feet. He won the long-distance record hands down, making a round trip of forty-five miles to the racetrack at Santa Anita.

One pilot's triumph was also a proud nation's sigh of relief. Since 1783, the French had been taking to the air in hot-air and hydrogen-lifted balloons. These graceful craft were colorfully decorated and elegantly fitted out. As they lazily hovered over the boulevards of Paris, it seemed logical that the self-appointed artistic, intellectual, and scientific capital of Europe would become the aviation capital of the world. In 1906 a certain Alberto Santos-Dumont built an aircraft that lifted briefly from the ground.

Technological innovations in early aviation were often introduced in flamboyant performances of skill, daring, and showmanship. Louis Paulhan, an experienced circus performer in France, became an air show superstar on both sides of the Atlantic. As photographers clicked their cameras before take-off, he ritually touched his wife's hand for good luck, underscoring the potential danger of his stunts.

But the Wrights' successes only two years later were a shock to Gallic complacency. In its first outing above European soil the American brothers' latest model soared over the heads of stunned French crowds for two hours and twenty minutes. No one in Europe had ever stayed aloft for more than an hour. Most flights subsided within minutes.

France woke up. In July 1909 Louis Blériot, who had amassed a fortune manufacturing headlights for automobiles, became the first aviator to fly across the English Channel, a distance of twenty-six and a half miles that he covered in thirty-seven minutes from Calais, France, to Dover, England. By then the Wright brothers had flown three times that far, but France was back in the running. One of Blériot's monoplanes used a configuration that is the basis for most light aircraft flying almost a century later. During the First World War he directed a factory that produced ten thousand aircraft for his country's military, including the fighter known as the SPAD. In aviation, as in many other realms of technology, national pride was to become a major spur to development.

Another was the growing influence of the mass media. At the L.A. competition

Paulhan shrewdly gave a brief plane ride to the influential, fabulously wealthy newspaper publisher William Randolph Hearst.

The Media Take Wing

"There is nothing with which to compare the sensation of flying. I felt that great sense of exhilaration which all aviators describe, and, in addition, a deep serenity—a calm enjoyment of what seemed to be the perfect condition of a new and better state." Hearst's reaction was of course not an impetuous outburst but a considered statement published in his newspapers. The publisher was on board in more than one sense.

Immediately he put up a fifty-thousand-dollar prize for the first aviator to fly across the United States in less than thirty days. Next, he ensured that newspaper coverage and attendant publicity for any attempts would be complete, fervent—and likely to have a healthy effect on the sales of Hearst papers.

Even the great showman himself could hardly have imagined a better contestant than the last to show up. In 1911 Calbraith Perry Rodgers, a six-foot-four former college

Far left: *Like the Wrights, Glenn Curtiss built bicycles before turning to aviation, but he was also a champion motorcycle racer and an inventor of engines. This contraption is a vehicle he built to test a system of controls he designed for a plane. On July 4, 1908, after building a new aircraft with the advice of Alexander Graham Bell and other experimenters, he became the first person to fly in public for as much as a mile.* Left: *The cover of a Parisian journal celebrates Louis Blériot's landmark flight across the English Channel in 1909. After years of experimentation and enormous personal expense, he was able to build the plane for this flight, the* Blériot XI *monoplane, only after learning how the Wrights used wing warping. The instant his plane touched British soil, the island nation lost her most important natural defense; enemy planes could now fly over the surrounding waters.*

football star, announced his intent to fly coast to coast in a Wright biplane. Controlling this particular craft in the air required lightning reflexes and brute strength, as Hearst reporters earnestly pointed out to the uninitiated. Rodgers's habit of clenching an unlit cigar between his teeth only added to the romantic image of the tough, brave loner. Hearing-impaired after a bout of scarlet fever, he had been unable to follow the naval career traditional in his family, and he was determined to make his own mark on history. He was good copy, in other words, but it is well to remember that he was not flying by flapping his arms. His plane was the latest machine developed by the hardworking Wright brothers and their workmen. It was dubbed the *Vin Fiz Flyer* in honor of a soft drink manufactured by the sponsor of Rodgers's attempt.

Nor did he exactly defy the elements alone. Cheered by thousands on his takeoff from Sheepshead Bay, New York, on September 17, he was followed by a train loaded with food, spare parts, his wife, his mother-in-law, and cases of Vin Fiz.

From the tabloids' point of view, this was dangerously tame stuff, but lucky Hearst got his money's worth. The second day out Rodgers clipped a tree and crashed into a chicken coop. That produced a deep gash on his impressive temple. On the next lap he landed in a potato field, breaking his landing gear. The *Vin Fiz Flyer* was rebuilt, then came close to being torn apart by souvenir hunters. Rodgers took off in haste, snagged a barbed-wire fence, and smashed to the ground again. Meanwhile his adventures were earning quite a bit of attention. As he passed over small towns, fire bells clanged to welcome him. Trying to avoid injuring a crowd of locals brought out by this clamor, he crashed yet again.

In short, it took him twenty-eight days to reach Chicago. To win the prize, Rodgers would have had to make California in two days, but he was undaunted: "Prize or no prize, I mean to get there."

In Texas the plane was demolished when Rodgers hit a stump. Meanwhile various parts had been popping out during flight, and he found himself flying blind through heavy curtains of rain. Four thousand feet above California's Salton Sea, a cylinder blew out, driving steel shards from the shattered engine into his arm. Three weeks after he flew out of Chicago and a mere twelve miles short of his goal on the Southern California shore, his reconditioned engine failed over the town of Compton, and he plummeted to a plowed field, suffering a concussion, a broken ankle, and various internal injuries.

Four weeks later, in a plane that retained only the original rudder and oil drip

An improved Wright aircraft was used by Calbraith Perry Rodgers, the first pilot to fly from coast to coast in the United States. Rodgers, who looked and acted the part of the sturdy adventurer, experienced many minor mechanical problems and five serious crashes. He literally had to rise from a hospital bed to bring his plane to rest at last on a Southern California beach.

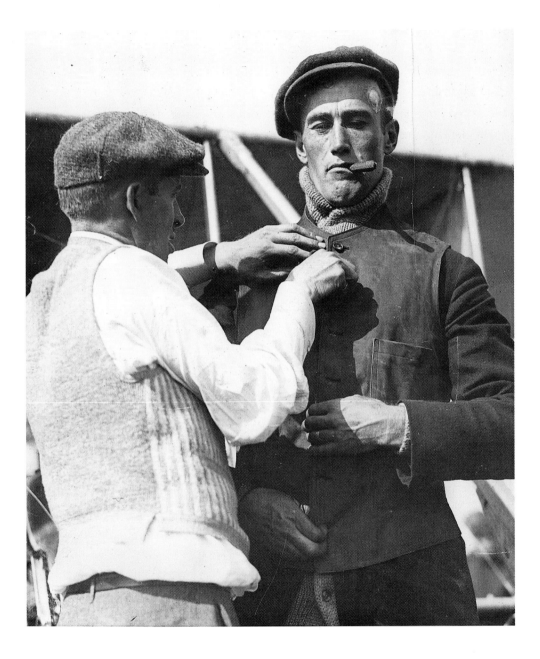

pan, with the pilot's crutches strapped to its wing, Rodgers put down on December 10 on the sands of Long Beach and let his wheels roll to the lip of the surf.

What the Accountant Sees

Reserved and stolid, Glenn Martin was an Iowan who had moved to Santa Ana and bought the local Ford dealership. There must have been a gleam in his eye that no one else saw, for each change in his life seemed to provoke the need for another. He was a good businessman with a keen eye for the bottom line. Unlike the more colorful aviators of the day, he steered wide of alcohol, tobacco, and women. He had earned an accounting degree on orders from his mother, with whom he lived.

Yet this quiet man fairly burst with enthusiasm for the products being designed and built by inventive teams of technologists and factory workers in those exciting times. It was oddly characteristic of him that he demonstrated the robust horsepower of Henry Ford's mass-produced motorcar by driving it straight up the steps of the Santa Ana courthouse.

Glenn Martin was determined to build on the success of the Wright brothers, but he was lucky to live through his first untutored try. After he instructed the mechanics at

his dealership to install a Ford engine on a glider built from a kit, the maiden flight of this monstrosity just missed cutting its inventor's head off.

Six years after the Wright brothers' famous flight, however, Martin and his crew did cobble together a workable motor-powered airplane. In large part they succeeded by borrowing ideas from a magazine article about a biplane built by Glenn Curtiss, the American sensation at the 1910 air show, but Martin had a different approach to the potential of flight: not daredevilry but entrepreneurship.

Eventually Martin became a regular on the aviation exhibition circuit, alighting at watermelon festivals and county fairs in highly unusual dress, a suit and shirt with a starched collar. Inevitably he was celebrated as "the flying dude." There was method behind this prim facade, for he was trolling for millionaire investors. He knew that planes were not toys and that aviation was not a circus act. He was a practical man. Meanwhile he plowed his stunt fees into developing new planes.

Speaking before a businessmen's club in 1911, Martin outlined the vision that drove him: "In the near future the airplane will become a thoroughly practical means of transportation for passengers and freight." This was, above all, a businessman speaking in the language of business to other businessmen. At the same time he continued barnstorming, attracting attention and some understanding of the potential he saw by ferrying mail, taking politicians and starlets aloft, and even doing a cameo role in a movie. But it was a tough battle: The railroads were working well enough for most down-to-earth commercial purposes, and their infrastructure was firmly in place. The same could not be said for airplanes.

The theatrics that finally paid off were inspired by the rumblings of World War I on the European continent. At the L.A. air show of 1912 Martin produced two elaborate demonstrations of aviation's future: a "mimic aerial battle" and a nighttime "battle-from-the-sky spectacle." His timing was impeccable.

A reporter for the Los Angeles *Express* immediately got the point: "[They] proved more conclusively than any other experiments ever tried that the aeroplane is practical in war."

Glenn L. Martin, one of the few early aviators with both a theatrical gift and a keen sense of the commercial potential of manned flight, demonstrated a kind of "air express" in 1912 by delivering copies of a newspaper printed in Fresno, California, to a city twenty-four miles away. This was also the year he startled an air show audience by staging mock battles between aircraft, thus accurately anticipating coming events in Europe.

The U.S. government caught on too. Backed by a contract from the War Department, the Glenn L. Martin Company was set up in 1913, charged with swiftly producing an armored biplane in a sixteen-thousand-square-foot assembly building in downtown L.A.

War in the Air

When 128 American citizens died in Germany's sinking of the British liner *Lusitania* on May 7, 1915, the outraged calls for strengthening the national armaments meant that even more national government support would be channeled toward aviation.

But how could the aircraft needed for serious involvement in a "world war" be produced quickly, efficiently, reliably? Already Europe had found part of the answer. In one country after another, even in Russia, sophisticated flight labs had been set up specifically for aerodynamic research. In German universities, for example, scientists were investigating wing theory.

These and other theoretical activities had been receiving real-life tests on the battlefield since August 1914, when World War I began. At first the airplane as weapon was little more than an accessory to the action. Field officers planning tactics sent a pilot up to spot targets for the artillery or to report back on the apparent movements of enemy forces.

But this change in intelligence-gathering capacity restructured the face of war. As accurately aimed mortar shells peppered troops, shooting up from hidden positions to blast coordinates determined by airborne observers, soldiers learned that they had to dig in or die. The result was a draining, static war, as combatants huddled in opposing trenches along a snaking four-hundred-mile-long line of terrified stalemate.

By contrast, the warriors of the air became more daring, more confident. At first the derring-do came from the British and French pilots, who tended to be upper middle class and well educated, with the penchant of the privileged classes for dangerous sport. They flew over the German lines to throw down bombs by hand, creating havoc. Then they began firing machine guns from the air as they flew. This first step in airborne ballistics was soon outstripped by a lethal German innovation, a machine gun synchronized to fire through the whirling propeller without hitting the blades.

In these mortally playful advances were born two new types of aircraft, sleek children of the exigencies of war: the heavy bomber aircraft, and the agile one-man fighter.

It soon followed that bombs ranged widely from the battlefield. In 1917 Germany began bombing civilians in London. There were few deaths, but the world was shocked, and not a little awed, that technology had lured seemingly civilized political and military leaders into such morally offensive acts. The marvel of flight became, at least briefly, the horror of aerial bombardment.

Yet the very lethal extension of warfare into the air at high speeds somehow took on the aura of romance. Pilots fought to the death in their swift, graceful craft. Their combats were linked in the popular imagination with the courtly jousts of medieval times. Death could rain down from the sky on infantry and townspeople alike in cruel bombardments. Still, the "birdman" looping his tiny craft into the clouds and beyond was admired as a new kind of hero in an increasingly troubled age.

Faced with the need to catch up quickly, the U.S. Aircraft Production Board decided to manufacture warplanes based only upon designs that had already been proved to work in the European conflict. Each would be supplied with the so-called Liberty engine, which had been created and was to be mass-produced in America using assembly-line principles developed by automaker Henry Ford. Indeed an enthusiastic Congress promised its allies in Europe a "Yankee punch" of twenty-five thousand new airplanes.

When America became fully engaged in the war, young Donald Douglas left the Glenn Martin Company to head the engineering department of the Army's Signal Corps, where he was to work with the European designs. He was responsible for America's only original airplane design in the war, the twin-engine MB-1 fighter-bomber. Although not a huge breakthrough because of its striking resemblance to the earlier British de Havilland, it was impressively armed with five machine guns and could carry a thousand pounds of explosives.

Meanwhile, as the carnage of the great conflict intensified, such effective technological developments of warfare as poison gas, armored tanks, killer submarines, and the airplane appalled the public and called into question the meaning and impact of science and engineering.

The war brought home a critical lesson about technology: A remarkable invention that can greatly improve daily life can also be an awesome engine of destruction and death. A technological wonder is no more than what humankind makes of it. Its meaning is created and affirmed by the ways in which it is used.

Immediately after the war the industry was dead in the water. The "Yankee punch" had had little discernible impact, since only 213 planes were delivered to Europe. Hundreds of thousands of government dollars had been spent too quickly in the rush to get U.S. military aircraft into the air. Corruption had been endemic. The assembly lines had in no way set high standards; in truth American planes were uninspired copies messily cranked out.

After the war the Glenn Martin Company managed to assemble only one aircraft a month. No one, including the Wrights, was doing innovative work in the United States. The public, dismayed by the many deaths of stunt fliers and famous aviators trying for new records, was not crying out for the development of air travel. Indeed, virtually no one within or outside aviation seriously considered the possibility.

But the most important failure was a matter of priorities. The United States, the center of aviation technology since the beginning of the century, had fallen far behind Europe because the government had not consistently invested in the field, as European governments had been doing.

To survive, aviation technology would have to prove itself. Bank accounts fattened when trains wove a commercial tapestry across America, stimulating new towns and businesses. Could the airplane provide similar growth all around? For the moment other technologies took the lead.

Ford's Answer

By 1915, the legendary Henry Ford was turning out inexpensive Model T automobiles at a fantastic rate, but it was the method more than the product that became a benchmark in the story of technology.

"I will build a motor car for the great multitude," Ford had announced in 1908. Previously the automobile had been a prestige toy for the wealthy. Beginning with his own fertile inventiveness and mechanical aptitude, Ford knew how to bring together other self-taught mechanics into a creative team, then set them loose to solve problems however they chose. Within little more than a year of hectic creative teamwork, Ford and his colleagues produced an automobile that was at once the simplest and most sophisticated ever built.

The fresh developments included a strong four-cylinder power unit, a semiautomatic planetary transmission, and a magneto instead of heavy dry storage batteries. The magneto, which used magnets to generate electricity, supplied ignition to the engine. In addition, for the first time lightweight steel casings enclosed the car's transmission, axles, and general workings, protecting them from rain, dust, and accidents. The Model T ensured that the gasoline-powered internal-combustion engine would overpower the day's alternative auto engines—electric motors and steam engines.

Ford's fundamental achievements, however, were more abstract. Like Edison, he knew that a talented, motivated team of creative individuals can create unprecedented solutions to knotty problems. His Model T also exemplified the virtue of simplicity. It could travel long distances with few repairs and relatively little maintenance. This remarkable car, so beloved by owners that it was called the tin lizzie and welcomed as a member of the family, was a perfect match with the needs of Ford's target consumers. Rarely do design and market environment overlie so completely.

The manufacturer's terse explanation changed complex industrial production forever: "The way to make automobiles is . . . to make them all alike . . . just as one pin is like another pin when it comes from a pin factory."

By 1917, the most powerful engine used in military aircraft battling in Europe was the four-hundred-horsepower American Liberty engine. German engines were less than half as powerful because of shortages of such materials as copper, fuel, and rubber. Developed by automakers in Detroit, the Liberty was so successful that it was the mainstay of the American military for the next decade.

Ford also saw that his profits depended upon producing as many cars as quickly and efficiently as possible. Parts had to be made so nearly alike that they were interchangeable and did not have to be hand-tooled to fit the individual car. Producing such parts required nothing less than a revolution in the development of machine tools, which had to grind or mill a part accurately to within one ten-thousandth of an inch. Increases in strength as well as accuracy were required, since metal cutting metal was authentically a violent act and the tools had to endure. Workers made the parts and fitted them together in subassemblies. These were then sent to workstations where a team built the whole car, moving from one task to the next. This concept of four subassemblies—frame, front axle, rear axle, power unit—was the innovation that put Ford products at the top of the market.

This newfound efficiency was still not enough, even though Ford's annual pro-

duction had doubled in 1912 to the astonishing total of seventy-eight thousand cars, a volume no one had ever thought possible. Attempts were made to set up an assembly line to produce the magneto. After several false starts the idea worked, with each worker contributing one part to the assembly as it moved along at a set rate. When the production of magnetos quadrupled, the heads of the engine and transmission departments demanded assembly lines too, and eventually the chassis was also assembled by the new method. The assembly time for a chassis dropped from four hours to ninety-three minutes.

In short, the Ford plant became a rapidly operating, smoothly running factory machine. It required fewer workers, even as production nearly doubled yet again, and the role of the assembly worker required little skill, thanks to special-purpose machine tools. There was little call for individual human involvement, but the dogged labor required was incredibly draining. Workers quit in droves, having lasted only four months on average. In response Ford added a very human factor: an unprecedented doubling of pay to five dollars a day. As the jobs became more financially attractive, however, there was more competition for them and little job security. It was the ethos of the assembly line that the well-machined worker, like a worn-out part, could be readily replaced. Moreover, good workers, in Ford's view, were to fit a mold: They should never touch alcohol, and they should be seen in church every Sunday. To enforce these convictions, he established a "Sociology Department."

Surviving letters of new immigrants sum up the trade-off in words like these: "The work here is very hard. The life is very hard. But if I don't get injured and I don't get laid off, I'm making very good money."

The assembly line anticipated and provided a model for a mechanistic society: People may become interchangeable parts in industries spawned by technological progress. Was a mechanistic view of humans the necessary trade-off for the new luxury products? It was becoming clear that the social impact of technology would be complex.

If technology was not necessarily making the worker's heart sing during his shift,

In the second decade of the twentieth century, virtually everyone in America knew what this scene meant: Thanks to the assembly-line methods and other innovations of the tireless Henry Ford, a working family might be able to afford a Model T, his relatively cheap, reliable, and easily maintained vehicle that brought unprecedented freedom of movement. Ford designed and made his first gasoline-powered engine in 1893, his first car three years later; he set up the Ford Motor Company in 1903.

it was providing a means toward increased comfort. For the first time in any country ever, factory workers moved into the middle class. By 1924 Ford's techniques were so efficiently productive that a Model T could be sold for only $290, or a third of its price fourteen years earlier. Some Ford assembly-line workers could afford to buy the products they made.

The Model T made it past the factory gate, however, only because other forces in America had been contributing to the infrastructure of roadways, the distribution of fuel, the passage of traffic laws, the regulation of driver's licenses, and advertising. All of these factors combined to help make Henry Ford the richest man alive, for his company manufactured half of all the automobiles in the world.

Enthusiasm for the automobile encouraged the rapid building of roads, which in turn encouraged the development of new neighborhoods and towns. Now it became

easy to live miles from work; people gained greater independence in their personal lives. The old centers of power and those who thrived there—the established leaders of politics, business, and "society"—became less relevant.

In 1996 the social historian David Halberstam summed up this nationwide revolution in the individual's sense of opportunity: "There is this concept of freedom that you can reinvent yourself. You don't have to live in the same town as your parents, you don't have to live the same kind of life, you don't have to do the job that they had. You can go where your dream takes you. If you want to leave where you are, you get in the car and drive."

Communication Takes the Stage

Two decades into our century an adman wrote, "These are times when the earth revolves twice as fast upon its axis." His profession was one of the accelerating forces because advertising thrived with the prosperity of new technologies, especially those flooding into the home.

Washing machines, vacuum cleaners, irons, refrigerators: these were mass-produced on the new assembly lines of the booming interwar economy, and the homes of millions of families, including young couples fairly delirious with the fast pace of change in the 1920s, were rapidly electrified to keep them perking away.

The glamour item in the interwar surge of mass-produced technological wonders was a small piece of furniture that seemed pure magic to delighted child, yearning teenager, busy parents, and world-weary oldsters alike, the radio.

The basic concept behind this twentieth-century sensation first gained public attention in 1896, when British customs officials seized and destroyed a supposed bomb carried by the young Italian inventor Guglielmo Marconi. This was a peculiar failure in communication, for the mysterious electrical mechanism was his so-called wireless telegraph and Marconi had been invited to Britain by the government after Italian officials had declined to support his research.

Invented just the year before, the device could send electromagnetic waves

Ford and his workers continually reinvented their manufacturing methods, driven to become increasingly more efficient by the demand for the Model T. In 1913, the flywheel magneto became the first part to be put together on an assembly line that was kept moving, lowering assembly time from twenty minutes to only five. This was effectively the birth of mass production.

through the air as far as a mile. Others had known about these mysterious "air" waves, but Marconi was the first to intuit a practical use for them. By interrupting the wave, he was able to send signals from a transmitter to a receiver, even with a hillock in between. The addition of an elevated antenna and a ground at both the receiver and the antenna sent the signal about a mile and three quarters. Working with British government funding, he was able to send signals as far as nine miles by 1897.

Marconi well understood the commercial implications of his experiments. In 1910 a fleeing murderer was arrested aboard an ocean liner because of a telegraph alert across the waters, but an iceberg in the North Atlantic may have revealed radio's power to the world: When the unsinkable *Titanic* sank in 1912, hundreds of passengers were saved only because a Marconi employee, David Sarnoff, reportedly picked up radio distress messages and alerted ships in the area. These radio signals were transmitted by means of the vacuum tube, or thermionic valve, invented in 1904 by John Ambrose Fleming, an English electrical engineer.

As with aviation, the military strategists of all nations in World War I adopted and eagerly explored the potential of the new medium. Its effectiveness was improved

by the invention of the triode, or audion detector, by the American Lee De Forrest. The triode incorporates a control grid that allows it both to amplify the sound of the signal and to transmit it through the air. The assumption of the day was that radio would eventually develop into a wireless telephone for communication between individuals. Because the De Forrest patent was such a good amplifier, AT&T snapped it up. The triode fostered the rapid development of radio and also made television possible.

Thousands of trained radio operators returned to America after World War I. Many had scammed spare parts from the military's stock of Westinghouse and AT&T supplies and could piece together their own radios. Their loved ones were not amused. The batteries leaked all over the floor or carpet, and there was nothing of general interest crackling through the speakers. These primitive radios were usually banished to the garage.

Chance, playfulness, ingenuity, and big business—the combination to be seen again and again when a technology lunges forward in sophistication or commercial reach—then came together on the waves emanating from a garage in Pittsburgh, and the big business of broadcasting was conceived. Until then radio was a technology that lacked a commercial identity. What was to sell?

Westinghouse Company executive Harry Davis glimpsed a part of the future when he happened to hear about one of his radio engineers, Frank Conrad, talking to other radio hams. Set up in his garage, Conrad played a few records from his collection when he grew bored with chitchat, and more and more listeners tuned in just to hear the music.

Far left: *Seen here at age twenty-two, Guglielmo Marconi had just invented the first wireless telegraph receiver, but it would be five more years before the indefatigable experimenter made a device that brought him fame.* Left: *The triode or audion detector, invented by Lee De Forrest in 1907, was the first vacuum tube to amplify as well as transmit radio signals. It effectively made radio communication practical by strengthening very weak electronic signals and producing stronger transmissions.*

Westinghouse builds radios, Davis reasoned, and professional radio programming could lure larger audiences, thus selling more radios. KDKA, one of the first radio broadcast stations in America, began broadcasting from the Westinghouse roof in 1920. When Warren G. Harding's election as President was announced over the air before the morning newspaper arrived, the public was amazed. Within two years 499 more stations were in operation, and a hundred thousand radio sets had been sold.

These developments were positively exhilarating for many people in America and around the world. Overnight the home was in contact with a kaleidoscope of real-time experiences. The radio had moved full tilt from experiment to field telephone to hobbyist's delight to something huge, not quite comprehensible, and pregnant with idealistic possibilities.

Who was not bewitched as Caruso belted out his inimitably warm high Cs in New York and invisible waves somehow carried the glorious sound through the living

room walls and into the speakers of the little box with its glowing dial? Not to mention the World Series, and the tribulations of misunderstood soap opera heroines, the pounding hooves of a posse pursuing rustlers, and the sensuous violins of a nightclub orchestra late on Saturday night?

As historian Roland Marchand puts it today, "There was a sense that there were vast opportunities for human improvement. Radio would bring people in closer contact with each other, and they would understand each other better. The nation would be more unified. The world would be more unified."

But radio was preeminently a means of transmitting the message chosen by the sender. Businesses, churches, clubs, and even newspapers leaped into the new medium. Having your own radio station was free publicity. Even the smallest station was on the ground floor of . . . something.

Program content was not a burning concern. Listeners were happy to hear amateur performers who had walked off the street into a makeshift studio. It was even more of a treat to hear your neighbors crackling through the living room speaker. Part of the excitement was searching for the most distant radio signal. The thrill came from beating your best effort, not from hearing a particular program.

This novelty eventually began to pale. It became tiresome to hear a fabulously remote station suddenly strafed by static or find it fading off unaccountably into the void. Meanwhile some programs were insidiously building loyal audiences. Most notable

Far Left: *Children playing with a large, early radio were likely to be fascinated more by the mystery of radio itself than by its ability to provide content.* Left: *Radio engineer Frank Conrad inadvertently became the first disc jockey. Shown here with his hobbyist's setup in the garage at his Pittsburgh home, he spent many hours communicating with other enthusiasts and began airing phonograph records to liven up the airwaves.*

among these was the *Grand Ole Opry* in Nashville, Tennessee, with its country music stars and cornpone jokesters.

The best listening was available near the big cities, where large radio stations could afford to produce strong broadcast signals or pay for special events. Chicago's WGN, for example, amazed listeners by broadcasting the excitement of the Kentucky Derby live, piping it in from Louisville over telephone wires.

Few stations could afford such extravagance. Worse still, many were starting to lose money, even with the least expensive local programming. The economics did not make sense, yet a hungry public was beginning to demand more from its home entertainment center.

At this crucial juncture the same people who wrote seductive newspaper and magazine ads touting the wonders of broadcasting sensed a different kind of opportunity. Those families gathered misty-eyed around the wondrous radio represented a vast untouched pool of money. Print advertisements often suggested that glowing radio waves conveyed something uplifting, even ethereal through the air. Why could they not be used to convey sales pitches too?

Still, this was unexplored, possibly dangerous territory. The advertising aces worried that the resonant intimacy of the radio voice—this human presence talking personally in someone's home—could backfire if the wrong note was struck. Government officials were also wary; after all, the airwaves were considered public property to be regulated in the public interest. Herbert Hoover, secretary of commerce, spoke for many: "[It is] inconceivable that we should allow so great a possibility for service to be drowned in advertiser chatter." David Sarnoff, the former wireless operator, put it another way in 1926, when he became president of the first radio network, the National Broadcasting Company. He understood that the introduction of advertising, which could potentially reap enormous profits for his company, would have to be gradual and subtle: "The finest results have been with the most delicate announcement putting over the [brand] name."

A year later the first experiments with radio advertising were "informational programs." A spokesman for a meat company, say, provided cooking tips "for the housewife" without directly touting the virtues of his own product.

Next, the admen created variety programs that featured performers who agreed to change their names to promote the brands, such as the group that became the A&P Gypsies or the soprano aka Olive Palmer and the baritone aka Paul Oliver, warbling for

Palmolive soap. Finally, the "interwoven commercial" was pioneered on the Maxwell House coffee show, as the characters sipped their coffee and waxed nostalgic about the good old days at the Maxwell House hotel, and so forth.

These were fledgling efforts, however. Radio became a successful commercial medium for one reason above all: the introduction of talented celebrity performers to the airwaves. Straight from vaudeville, where they had earned national or even world-wide fame, and over the airwaves traveled the music and comedy of Eddie Cantor, Jack Benny, George Burns and Gracie Allen, Fanny Brice, and Rudy Vallee with his Connecticut Yankees. These entertainers not only drew huge audiences but were also clever about integrating the commercials into the flow of the program, making them acceptable and even entertaining. Unfortunately the success of certain performers cut down the variety of programs offered. Advertisers and programmers gravitated toward the proved, safe center in order to maximize the potential audience.

By 1927 the Federal Radio Commission, established to oversee the industry in the light of the "public interest, convenience, and necessity," determined that advertising was indeed in the public interest. A year later the American Tobacco Company experimentally suspended nearly all print advertising for Lucky Strike cigarettes and concentrated on broadcasting paid testimonials. Sales soared by 47 percent.

By 1935 the technology of radio had found its identity, a public medium dedi-

By 1928, when the orchestra for the Michelin Hour *debuted in costume as animated automobile tires, broadcasters and advertisers were clearly racking their brains to give radio entertainment a distinctive image. Surprisingly, the new medium would profit most from the talent developed in old-fashioned music halls and the vaudeville circuit.*

cated primarily to mass entertainment and financed by paid advertising. This combination of factors had never existed before. It became the definition of broadcasting. Later television virtually piggybacked the concept. Appropriately, when Marconi died in 1937, the world's broadcasters let the air fall silent for two minutes in his memory. It was a dramatic demonstration of how much radio noise had flooded the globe because of his life's work.

Wherever it was beamed, radio reflected the host society. The blend of advertising and entertainment that worked in market economies was considered dangerous by the oligarchy that ruled the Soviet Union. Citizens were given a *radio-tochka,* meaning "radio-point," which could receive only one frequency and upon which they heard nothing but the official government view of the world. It was not at all inevitable, in other words, that radio would develop as it had in the United States.

The Business of Film

Perhaps the most extraordinary coming together of technology, entertainment, and big business occurred in Southern California, where the powerful, if flickering, images of the so-called silver screen began to define new American fantasies. The infant movie industry incorporated three major forces essential to twentieth-century technological development. Like radio, the movies had to find a unique identity for the technology of film. Like the automobile giants, movie producers learned to develop assembly-line technologies. As in the aviation industry, competition spurred innovation.

Both the camera and the projector for motion pictures were first made workable in the 1880s at Thomas Edison's laboratory in Menlo Park, New Jersey, primarily through the ingenuity of his colleague, W.K.L. Dickson. In 1896 Dickson perfected the first workable camera/projector system. His kinetograph was a camera that rapidly captured a series of individual images on a continuous roll of film. His kinetoscope reeled this film, illuminating one frame at a time, so quickly that the human eye enjoyed the illusion of movement.

In France the Lumière brothers, among others, immediately grasped the poten-

The poster for the grand opening of Universal City stresses two points: the studio's superiority in size and resources for making motion pictures, and the diversity of the settings of its stories, from the Wild West to a Turkish bazaar. The fantasies of Hollywood moviemaking were grounded in business realities, including use of assembly-line techniques that mirrored those in Ford's plant in Dearborn, Michigan.

tial and first showed films to paying audiences in 1895, but Edison himself thought that coin-operated peep shows allowing only one viewer at a time would be the chief commercial future of film.

Scores of inventive, even daring creative artists and wildcat investors proved him wrong, but not right away. The first movies in the United States were shown in penny arcades, where a cent bought a few seconds' worth of action in a peep show. Even at this humble beginning, filmed entertainment immediately encountered the kind of moral outrage that bedeviled filmmakers throughout the century. Women, it turned out, lined up eagerly to watch boxing films at the penny arcade. Was this seemly—ladies viewing bare-chested men battering each other?

Then the first theatrical films earned greater profits than any other form of entertainment. Filmmakers tried various genres, from filmed stage plays to real-life documentaries, like a film about the 1906 San Francisco earthquake. But original stories conceived to exploit film's special techniques turned out to be the most popular with audiences and the most lucrative for producers.

In the first decade of the century the Southern California sunshine, its cheap and open land, and the opportunity to evade a motion-picture equipment monopoly in the East were beginning to attract filmmakers to the L.A. area. Dismissed as "the film colony," they and their sometimes disreputable hangers-on inspired such rental signs as NO DOGS OR ACTORS ALLOWED.

Not for long. By 1912 such cinematic legends-to-be as D. W. Griffith, Mack Sennett, and Cecil B. DeMille were in town. They and others were going to turn this infant technology into an instrument of high art . . . on occasion. It was also a technology that could manufacture product by Ford's assembly-line methods. Fictional films created at a studio lent themselves to a centralized production process. By 1920 the largest Hollywood studio, Carl Laemmle's Universal City, could churn out 250 serials, shorts, newsreels, and low-cost feature films every year.

Laemmle, who had got his start in the nickelodeon business, had the good sense or fortune to hire Irving Thalberg, a frail young native New Yorker with a chronic heart condition and solid gold intuition about the future of film. Dazzled by the potential of the huge Universal City studio—its film-processing labs and cutting rooms, construction yards and costume shops, back lots and shooting stages as large as football fields, restaurants and police force and zoo—Thalberg recognized at once how entertainment value, business sense, and technological innovation could be blended in a system unique to filmmaking.

Laemmle had always known that each product on the assembly line had to seem commercially different from all the others—what he called a "scientifically balanced program." On the other hand, following his ruling concept of "regulated difference," each genre was made within a rigid mold, once a story formula and production process had been firmly established. Young director John Ford, for example, could crank out five-reel westerns starring Harry Carey one after the next, minimally adjusting story details and even reusing action footage over and over. This Universal City product typically found its audience in rural and suburban theaters.

Thalberg provided a new vision. He saw that the future lay with top-quality, uniquely envisioned films featuring charismatic "stars" and created by gifted directors. He also realized, after watching the illustrious director Erich von Stroheim squander time and money recklessly on a film called *Foolish Wives,* that a balance had to be reached between the director's essential artistic control during shooting and the studio's need to consider the financial bottom line. Creative people might imagine themselves at a "studio," but the executives at the top knew they were running a factory.

Thalberg aimed to produce profitable excellence. In 1923 he switched to Metro-Goldwyn-Mayer, where a strong star system was being put in place. As vice president in charge of production he was assigned to keep these walking commodities busy and happy, and he understood their market value. But familiar screen presences, such as the legendary Lillian Gish, were fading. When Thalberg cast Greta Garbo as the sexy temptress in the silent film *Flesh and the Devil,* she sold out movie houses all over the world. The successful development of sound movies, or talkies, attracted new legions of fans. A new generation, including Clark Gable and Joan Crawford, began to shine from the world's screens. These "stars" were the basis of a system as rigorously calibrated as Henry Ford's.

Technological advances were artfully used by Thalberg to set up a production system that earned MGM record profits during the Depression while his competitors fell deep into the red. The details of each motion picture were carefully planned out on a daily schedule and closely supervised to ensure that every element was being well produced on time. Thalberg himself moved nonstop during the day from a script conference for the outlining of one movie to a set and lighting check for a work in progress and on to an editing session for a movie in postproduction. Each new technological development, from sound-on-film to sharper-focus lenses to new lighting techniques, was effortlessly melded into his system.

Meanwhile increasingly adroit artistic teams were creating images of light, symphonies of sound, that profoundly affected the national psyche.

Garbo was only one of many screen legends to suggest a powerful, enviable personality while ostensibly losing herself in a role. Never before had human beings seemed so luminously beautiful, so grandly sensual. Garbo and others might play queens, spies, or hatcheck girls, but audiences learned new ways of staging their own lives from these

A postcard city come to life as a sound stage, the Monte Carlo set for a forgotten film, Foolish Wives, *was meticulously detailed to look as real as the city itself, but director Erich von Stroheim's artistic integrity cut deeply into the Universal studio's operating budget. At MGM, Irving Thalberg avoided such expensive mistakes by combining advancing technology with strict cost-control guidelines, earning record profits for the studio even during the depths of the Depression.*

charismatic stars: how to dress, how to attract the opposite sex, how to tough out adversity, how to take a dream by the scruff of the neck and shake out a new future. Studios, for their part, learned to build their total marketing strategy around their stars and promote new ones with reliable success.

Educators and religious leaders and politicians wondered about the moral effects of this compelling technology. Was a particular film "true"? Would it corrupt? Did it inspire?

The Reflection Reflected

Easier to answer were questions about the interaction of the arts and technology. In *Grand Hotel* and other movies of the late 1920s, set designers worked in the style known as art deco. Its straight lines, streamlined curves, slender geometric forms, and planes of contrasting color paid explicit homage to the latest technology, echoing the simplicity of mass-produced objects and the sweep of flight aerodynamics.

Popularized by the movies, this design style became synonymous with modernity throughout American life, and modernity in this country in this century has generally been synonymous with optimism. In 1930, as the Depression began to take hold, the completion of the art deco Chrysler Building in Manhattan was touted as a "knockout for pessimism." Its stylized eagle gargoyles, sculpted to soar, its nickel-plated radiator caps, indeed its upward sweep of line vaulting into the sky provided gleaming hope in a sullen age.

Not incidentally the building had been financed by a company that had been founded only five years before but was already one of the dominant automobile companies in America. Secretly the Chrysler's builders assembled a final spire that surprised the world. Once it was fixed on top, the architectural masterpiece became a statistical wonder, the tallest structure on the planet at 1,048 feet. A few blocks to the southwest there was disappointment all around until someone devised a similar tactic. A mast ensured that the new Empire State Building would be completed at a height two hundred feet superior to the Chrysler.

The essential engineering technology for both these impressive structures—a lightweight skin over a steel supporting grid—had been in use since the turn of the century, but never before had it raised skyscrapers with such efficiency, speed, and economy of effort. In the spirit of Henry Ford, the builders of the Empire State Building put together subassemblies elsewhere and brought the finished parts to the site; precisely when needed, a part was put into one of several specially built construction elevators and lifted to the appropriate place. Construction of the 102-story building took only a year, with a new floor going up every day at the peak of the building process.

Hollywood paid heed. Two years after it stood complete, the Empire State Building became the site of King Kong's last stand. The powerful creature of untamed nature, mesmerized by love, could choose no symbol more fitting in his struggle against technological civilization, even as airplanes circled like bees.

In real life these modernistic spires attracted people and machines; moreover, they were serviced by a complex system of public transportation that exacerbated urban congestion. Technology's achievements inevitably presented new problems to be addressed by newer, more creative technology.

Imaging the Future

While the movies sparkled during the Depression years, daily life became a deadening grind for most. At the celebrated 1939 World's Fair in Queens, New York, the theme "Building the World of Tomorrow" became an audacious challenge. In reality the present was bookended by the past ten years of poverty in the Depression and the terrifying early indications that another world war was inevitable, a war that would stimulate and be stimulated by technology. But on the thousand acres of glittering fairgrounds one could believe in a future beyond poverty and war, a world of material plenty made possible through carefully planned use of technology. This was not the world of the radar and rockets that were to come soon but the world of nylon stockings and fluorescent lighting, of television and faster, more powerful automobiles.

In fact, the first news event ever carried live in the United States on television—a canny blend of news and promotion—was David Sarnoff's dedication of RCA's exhibit. He spoke of the new technology of television with the high-flown rhetoric familiar from the early days of other potentially commercial wonders of the century: "It is with a feeling of humbleness that I come to this moment of announcing the birth in this country of a new art so important in its implications that it is bound to affect all society." The TV set on display had a seven-inch picture tube, but the image was enhanced, as the

Completed in 1930, the Chrysler Building in Manhattan is an architectural masterpiece of a style known as art deco, which became popular at a Parisian fair five years before. Sleek, abstract, dynamically curved but also dramatically angular, the style was influenced by the strong lighting contrasts of black-and-white movies. The building's piercing spire, covered with a blend of nickel steel and chrome in honor of Chrysler's cars, is a single section weighing twenty-seven tons.

phrase went, with mirrors and enthroned in a large wooden cabinet. Television was featured at other company pavilions as well, including Westinghouse, General Electric, and even the Ford Motor Company.

These were wonders indeed, but reporters looked for evidence that humans were still themselves. At the AT&T pavilion a limited number of fairgoers were chosen to make free long-distance calls to exotic locales while others eavesdropped on earphones. *Time* magazine reported one embarrassing result, which suggested that technology is always a little unpredictable. One woman called her kid brother as her boyfriend and other attendees listened. "Hi, screwball," said the boy. "Have you hooked him yet?"

At the Westinghouse building a robot straight out of science fiction performed simple tasks, but most adult fairgoers were much more excited by the electric washing machine on display. Still, novelties made a splash. The Ford pavilion highlighted the so-called Novachord, an electronic synthesizer. It used vacuum tubes to produce a fascinating range of sounds that aped acoustical instruments: piano, harpsichord, trumpet, guitar, violin.

A young boy, famous later in the century as the astronomer Carl Sagan, was awestruck to "see sound," as an oscilloscope screen displayed a sine wave in response to the striking of a tuning fork, and to "hear light," when static sputtered from a photocell activated by the beam of a flashlight. "Plainly," he wrote decades later, "the world held wonders of a kind I had never guessed."

The hit of the fair, brainchild of the imaginative theatrical designer Norman Bel Geddes, was an immense animated display called Futurama. Incorporating several million structures over 35,738 square feet, the model was funded by General Motors as an answer to increasing public concern about the new urban quagmires created by the interaction of traffic congestion, narrow city streets, and skyscrapers. General Motors was also on board, having been assured by Bel Geddes that visitors would be persuaded that the company was concerned about traffic safety, that the highway system should be greatly expanded, that automobiles would always be a fine thing to buy.

The famous symbols of the 1939 New York World's Fair, the needle-like Trylon and the Perisphere, marked the entrance to a thousand acres of the future. After a decade of uncertainty as world economic depression deepened, exhibitors heartened fairgoers with promises of a life made more comfortable, convenient, and entertaining by emerging technologies.

Walking up a dramatically winding ramp through the doors of an intentionally unprepossessing building, the public first encountered a map of the United States with the 1930 highway system outlined in red electric bands. In a flash the lighting became a jumble of constricted lines, an image of the traffic congestion projected for 1960 if things were not remedied.

But solutions lay ahead. Seated in plush easy chairs attached to a moving cable, visitors found themselves seeming to fly low over a fourteen-lane superhighway of the future. There were no roadside buildings or annoying intersections, just a ribbon that flowed continuously as the traffic exited and entered by means of cloverleafs and ramps.

Where were the people? Thanks to the automobile and the freeways that facilitated its movement in 1960, Americans could now live in the "middle landscape," a dream of visionaries since the nineteenth century. The ranks of automated easy chairs glided over precisely detailed miniature models of widely scattered recreation areas and farms, dams and hydroelectric plants, even orchards with trees under individual protective glass domes. Then there were carefully planned communities where technology had harmoniously integrated work, leisure, and domestic life, with clean up-to-date industrial districts to the side. There were no models of slums or billboard clutter, for this was, as a disembodied recorded voice assured the fairgoers, "the world of 1960."

Historian John Staudenmaier, considering the Bel Geddes exhibit from the perspective of the 1990s, places it in the context of the singular American experience: "There is a mythology that says that Americans didn't come to an occupied country and then overpower and conquer it. We came to an empty land with a clean slate. We started to create from scratch. That's a powerful American technological dream. Wouldn't it be great if we could just sort of start fresh?"

The fair was a moment for disregarding the immediate future, an exuberant manifestation of the faith that science and technology would inevitably create a better world.

Had there been a companion Futurama dedicated to the international cataclysm on the horizon, its theme might have been something like this: "Welcome to tomorrow . . .

Seen holding plans at upper left, the celebrated Broadway set designer Norman Bel Geddes looks over the Futurama exhibit at the New York World's Fair. His quarter-mile-long scale model, called "the little world of the future," portrayed a synthesis of technology and humane concerns with a clear messsage: The automobile and other technological innovations can be controlled so that congestion, pollution, and poverty will vanish.

when technology will be able to destroy the entire world, even the future itself." World War II was to thrive on technological breakthroughs, and vice versa.

The Watershed

A major aspect of the World War II years in America was prefigured—though who could have known it?—by the sheer nylon stockings that were enthusiastically promoted at the New York exhibition as one of the century's major miracles.

Introduced to the public by E. I. du Pont de Nemours & Company, nylon was described as "the first man-made organic textile fiber." No communications medium ignored this story; from "coal, water and air," according to a company spokesperson, technology had created a synthetic thermoplastic material that could be "fashioned in filaments as strong as steel, as fine as the spider's web." The first supplies were sold out before Christmas. The new stockings were vulnerable to snags and runs, but their elasticity and overall strength were generally considered superior to silk.

Once again a seemingly humble product, part domestic convenience and part sexy glamour, had unpredictable and finally tremendous resonance on the world stage.

From the 1930s through World War II and into the Cold War was the unquestioned golden age of twentieth-century technology. Du Pont, which had become very successful over its two-hundred-year history by buying up patents for materials like cellophane developed outside the company, had anticipated change as early as 1927. That year, Charles Stine, the director of the chemical department at Du Pont headquarters in Wilmington, Delaware, set up the first corporate laboratory devoted to pure research. The best new ideas, he believed, and the bright young scientists who would discover them could be encouraged to thrive in an atmosphere of subsidized, unrestrained experimentation.

Some critics dismissed Stine's bold vision as "Purity Hall." On the other hand, leery scientists suspected that the company's profit mentality would insidiously skew their work in the laboratory. Many refused to become involved.

But in 1928 a thirty-one-year-old Harvard instructor, Wallace Hume Carothers, took the plunge. Uninspired by teaching, passionate for research, he was finally lured by Stine's guarantees of unrestricted freedom and the right to publish his findings. Although subject to occasional nadirs of depression, which he called "neurotic spells of diminished capacity," this polymath flourished. He combined learned interests in art, sports, politics, and music with an unusual gift for friendship. His colleagues loved to hear him talk.

"Harvard was never like this," he wrote. "I occupy myself by thinking, smoking, reading, and talking. . . ." Freed from the distractions of daily life, he and his team worked happily in a quiet lab in woods beside the idyllic Brandywine River.

This enviable life took place at a time when natural fibers and organic materials prevailed. As in biblical days, most clothing was woven from cotton, wool, or linen. Furniture was made of wood, and cooking pots and pans were heavy cast iron. But this traditional world was inefficient, often uncomfortable, and less visually appealing than the uncanny new world that began to take shape in "Purity Hall."

At Stine's suggestion, young Carothers began studying polymers, a name constructed from two Greek roots: *poly* for "many," *meros* for "part." Long and stringy in the shape of networks or chains, these molecules can be composed of as many as one hundred thousand repeating units. Whether human-made or natural, organic or inorganic, polymers have many attractive characteristics, including the smoothness of silk, the resiliency of wood, and the bounce of rubber.

Carothers and his associate Julian W. Hill put together a team to ascertain the structure of polymers by heating the individual units, or monomers, which then reacted with each other to form new polymer chains, link by link. Some of the chains they produced were so long that Carothers dubbed them superpolymers, but they were not long enough for him. Meanwhile he wrote a series of scientific papers that became the groundwork for all future study of these remarkable molecules. The goal, which proved elusive for months, was to build a polymer molecule with a chemical weight greater than 4,200.

Meanwhile, one of Carothers's assistants, working under his directions, discovered a process that scientists had been seeking for some seventy years. On April 17, 1930, Arnold Collins accidentally took the first step toward paying off Du Pont's lab costs, and exponentially more. As he experimentally polymerized an unusual compound called DVA, Collins saw emerging a new white liquid compound that seemed to have something solid inside. Puzzled, he squeezed this material as if it were a sponge. With the liquid pushed out, it solidified into white, rubberlike masses that could be bounced across a tabletop. More important, they sprang back into their original shapes when deformed. Neoprene, the first synthetic rubber, was to prove sturdier than nature's version and became critically important in World War II.

From Laboratory to Marketplace

Soon Carothers and Hill chalked up another world-changing discovery. Carothers made an inspired guess about their failure to construct a molecule with a chemical weight greater than 4,200: The reaction between the acid and alcohol used in their experiments produced water, which halted the chemical reaction when it accumulated. He ordered a special still that could extract the very last water molecule out of the experimental materials. Hill left the next test substance in this still for days, just to be certain. When he dipped a glass stirring rod into his molten polymer compounds, he was able to draw out a thin, "taffylike" filament. When it hardened, this filament could be stretched to four times its original length, and then, in Hill's words, "suddenly, any further pulling simply cuts your fingers. You have the sensation of virtually feeling the molecules lock into place." Memorably Hill referred to his discovery as a "festoon of fibers." Not only did the new polymeride retain its new length, but it also demonstrated extraordinary flexibility and strength. The exhilarated scientists took turns racing down the halls, stretching the fibers to extraordinary lengths. Nature's fibers did not behave this way. Were we discovering . . . or were we creating? In any event, Du Pont's researchers now understood the process of building polymers.

The search for a viable commercial product was far from over. Because their synthetic fiber had a low melting point, it did not survive the laundry tub or the ironing board. Company officials became impatient; there was too much delay "between the test tube and the counter." While the rest of the company lost a third of its workers to the Depression, the chemical department had enjoyed special protection—but no longer. Elmer K. Bolton, who originally opposed the pure research program, took over direction of Du Pont's lab.

Bolton believed that chemists should tackle specific, well-defined problems, not engage in undirected research. It was, after all, the company's tab. At his insistence Carothers returned to nylon experiments. It hardly helped that this sensitive individual

Top right: *This false-color computer illustration of a section of Nylon 66, the form of nylon that transformed the clothing industry and saved lives in World War II, is coded to indicate the different atoms in the polymer chain: green for carbon, white for hydrogen, blue for nitrogen, and red for oxygen. Bottom right: Wallace Carothers, shown at work in the Du Pont company's experimental station, directed the research on plastic polymers that resulted in the invention of nylon in 1934.*

had to struggle against both his own insecurities about his ability to produce practical technology and his recurring depression. Even as he tried strenuous exercise to raise his spirits, he carried with him at all times a small capsule of cyanide, a flirtation with suicide.

In May 1934, after two months of intensive work under his supervision, one of Carothers's assistants drew a commercially viable fiber from a four-gram batch of amide polymer. Du Pont initiated a crash program dedicated to this substance. Two years later Carothers was elected to the National Academy of Sciences and universally acclaimed as the world's foremost polymer chemist, but his true legacy was a world virtually remade by synthetic fibers.

Meanwhile the Du Pont labs discovered nylon's seemingly endless versatility, processing it into filaments, yarns, coatings, films, and plastics. So far, nylon had been a secret well kept—and not by accident. After the first test of its knitting potential in February 1937, every stray bit of yarn was collected and weighed to ensure that none left the premises. Mechanical knitting machines were set up to produce miles of hose. Each of these mechanical weavers could knit hosiery from a single strand of nylon fiber.

How could this "miracle fiber" be most unforgettably introduced to the public? Du Pont shrewdly concentrated on nylon stockings, revealing the product to an amazed public on October 27, 1938. The copywriter's comparison with alchemy—for the lovely silklike hosiery had indeed been transmuted from coal, air, and water—was no exaggeration. This was technological sorcery. The excitement was still feverish a year later at the 1939 world's fair. One woman's comment spoke for most: "My legs feel different. I feel like kicking them up in the air."

By chance the nylon industry unpredictably sparked the growth of a related industry. Previously ladies' stockings had been made of fabrics—silk, cotton, lisle, rayon— that were not sheer and revealing. Women were not in the habit of shaving their legs. Now, thanks to nylon, there was a growth surge in the razor industry.

But the chance necessities of history probed other properties of nylon, and lives were saved during a new global conflict. Tougher than silk, nylon made for much more reliable military parachutes; besides, silk imports from Asia were cut off after the Japanese bombed Pearl Harbor in December 1941.

The curious synthetic fiber extruded by accident in "Purity Hall," as one man pursued his own research despite private demons, was overnight declared a strategic commodity. Nylon stockings were banned as the fiber was used to manufacture airplane tires, flak jackets, gunpowder bags, and numerous kinds of ropes for the war effort. In a door-to-door effort throughout America, millions of women contributed their stockings for recycling as parachute yarn: "Taking 'Em Off for Uncle Sam."

War to Peace

The war's accelerating hunger for bombs, weaponry, ships, and tanks quickly exhausted supplies of raw metal. But polymer-based plastic goods rolled off assembly lines into the breach: machine-gun parts, bayonet scabbards, and even bugles. Thanks to layers of polymer resins, plywood gained new respectability as a structurally sound material useful in the theater of war in aircraft wings, gun turrets, and the hulls of PT boats. Back home in America, when supplies of wood were used up, newspapers and wood pulp

During World War II, the nylon made familiar as ladies' stockings became essential to national defense in the form of parachutes, ropes, tents, flak jackets, and aircraft tires, among other uses. The crowd thronging the entrance to a San Francisco department store has been attracted by the first sale of nylon stockings at war's end. Postwar consumers would soon have a wealth of products to buy, thanks to mass production and rising incomes.

were bound with polymers to make construction board, the basic building material for thousands of new factories and worker communities.

But other industries changed too as domestic factories became essential parts in a hyperactive war machine. The fertile marriage of war needs and government money produced the most spectacular technological results in Southern California. Because of its strong history of aircraft production and its proximity to the Pacific conflict, technology industries there received more of Uncle Sam's dollars than factories in any other area of the country.

From 1940 to 1947 more than five hundred thousand people moved to the L.A. area to take jobs in defense industries. Because the military had drafted so many white males, many of these workers came from groups formerly considered "unacceptable" by U.S. industry: African Americans and women. South of L.A. in Long Beach, government funding made it possible for the Douglas Aircraft Company to build one of the world's largest, most advanced aircraft plants. Round-the-clock production there eventually employed a work force of forty-three thousand. As one result, the nation's demographics were permanently altered; the population of the West Coast increased by almost 40 percent.

Rushing to meet the challenges of war, the government-financed California building boom created a huge system of large aeronautics factories integrated with the various specialized small producers that supplied their needs. Never before had so much technological innovation been concentrated in one place. With blinding speed here and throughout the national war economy in the mid-1940s, the reveries of science fiction became concrete fact: jet-propelled aircraft, rockets, radar, synthetic high-octane fuels, and the weapon that seemed to fold time upon itself, the atomic bomb.

In a sense World War II was a huge technological experiment, marred by the disasters of failed equipment and lost battles and wasted lives. Yet many believed that the failures were eminently worth the gains of American progress in the long run. In search of new and better answers, firm connections were forged among the military, industry, and academia, a structure of intercommunication and influence that lasted for decades. At war's end the future looked bright with immeasurable potential.

At a glance it seemed that everyone's income was steadily rising some 10 percent each year in the postwar economy. Young couples raised on farms without running water or in cold-water tenements during the Depression were now able to buy tract houses. William Levitt, to take one example, was constructing entire neighborhoods almost overnight, using Ford's techniques of mass production.

New highways were being built, especially in the technologically superheated economy of Southern California. Engineers used banked curves, graded straightaways, and cloverleaf interchanges to create a road system that was essentially an engineered machine itself. The freeways astonished drivers used to narrow roads with many stops and poky traffic. Now it was possible to drive for miles without stopping or even slowing down.

In 1956 Congress authorized the construction of forty thousand miles of national highway, a typical example of government support of technology to produce social change. Already more than three quarters of all new housing was being built in the suburbs made possible by the automobile. At the same time this was a revolution in living for the white middle class alone. Policies of the Federal Housing Administration made it easier for them than for African Americans and other minority groups to receive housing loans and mortgage insurance. Racially mixed neighborhoods were bureaucratically discouraged.

But for those privileged to enter the suburban American dream, life was more comfortable than for the middle class anywhere else in the world. Technological innovations available only to the wealthy at the beginning of the century were standard equipment for the contractor-built tract house: central heating, indoor plumbing, telephones, washing machines, refrigerators, and automatic stoves.

Imaging with Numbers

In 1943, when aircraft increasingly became more complicated and pilot training more challenging, the U.S. Navy sought help from MIT's Servomechanisms Laboratory: Could a flight simulator be built both to speed up pilot training and to evaluate experimental aircraft designs before they were built? It would have to be extremely flexible in order to adjust to the testing of many different designs. Lab director Gordon Brown gave this task to Jay Forrester, a brilliant young graduate student, because the tall, naturally authoritative midwesterner had a rare aptitude for mathematics.

A simulator is a mechanism activated by mathematical equations that convey information about the force of engines, air and gravity upon the imagined aircraft. In sum, a simulator is a computer. In order to eliminate any lag between the trainee pilot's actions and the simulator's response, hundreds of differential equations have to be carried out at the same time. Unfortunately the day's existing analog, electromechanical gears, and cams were not sufficiently accurate. Jay Forrester did not find a solution by war's end, but the Navy was still interested in his work, now funded under a new name, Whirlwind.

Meanwhile, in 1944, Howard Aiken had created an electromechanical calculator at Harvard College with support from IBM. This Automatic Sequence Controlled Calculator, known as Mark 1, was eight feet high and about fifty feet long. It could add or subtract numbers in three tenths of a second, but division took twelve seconds. Forrester believed that its gears and levers were much too slow for his purposes.

In 1945 he visited another computer project, the Electronic Numerical Integrator and Computer (ENIAC), supported by the Army at the University of Pennsylvania's Moore School of Engineering. It had not been completed in time to fulfill its war mission, which was to speed up the calculation of the complicated ballistics tables used to aim weaponry. Still, it could compute complex equations with a rapidity that drew the interest of scientists around the country.

But ENIAC's vacuum tubes also created terrific problems. When the computer was turned on, its eighteen thousand tubes reportedly dimmed lights all over eastern Pennsylvania while producing a great deal of heat. A few failed tubes could bring the whole computer down. Some two thousand had to be replaced each month during full-time operation. A final problem: To change the programs that controlled the computer's operations, technicians had to spend hours moving about plug wires on the front of the machine.

As a first step toward the goals of Whirlwind, Forrester was able to improve vacuum tube reliability significantly. Then he greatly simplified and speeded up his device by converting his computations from numbers into a special code. The digital number system, or binary arithmetic, is based upon only two numbers rather than the ten of our decimal system. In other words, any number can be digitally represented as a combination of ones and zeros. Electronically these two possibilities are produced by switching from on to off. To take a simple example, 1101 is a digital representation of the number 13. When you count from the right, the value of each num-

Far left: *ENIAC, put together at the Moore School of Engineering in Philadelphia in 1945, was a behemoth that spread over a thousand square feet of floor space, but it revealed the potential versatility of computing devices. It was a visit to ENIAC that inspired the work of John von Neumann, the Hungarian-born mathematician who developed the use of binary numbers rather than the decimal system for information processing.* Left: *Jay Forrester examines a memory storage tube from an early version of Whirlwind, the electronic computer developed in the mid-1940s to make flight simulators more accurately indicative of flying challenges presented by different types of aircraft in a variety of situations. Whirlwind was the first real-time computer to use the devices that led to the design of today's video display terminals, and played an important role in civilian air traffic control and U.S. air defense.*

ber is doubled—hence, the base two—so that in this instance 1 is added to 0 to 4 to 8.

But few of the necessary people were impressed by Forrester's advances. For example, some of his academic colleagues described the project as an overblown boondoggle; the Navy, which was spending a tenth of its entire research budget on Whirlwind, was about to pull the plug. Forrester was frustrated because the nature of the work as basic engineering was not understood by the cost-accounting government overseers: "They miss what is happening. They miss the fact that you really have to go back into some very fundamental research and work right from the ground up to redefine electrical engineering in order to build digital computers."

Then, with one tremendous explosion in 1949, the U.S. academic/ government/industry effort was supercharged to build efficient computers as quickly as possible. National security was at stake. The Soviet Union's first test firing of an atomic device produced a devastating fallout on U.S. national pride and political goals. Suddenly the great, bustling, technologically sophisticated urban centers from coast to coast lay vulnerable to near-instantaneous destruction. It could come down from the very heavens that America's aviators had apparently conquered mere decades before. The vast oceans to east and west no longer offered protection from an enemy's destructive might.

Some military analysts believed that the only possible defense against nuclear attack was a preemptive strike. But President Harry Truman, along with many scientists, believed that technological defenses could be developed. The Air Force was ordered to come up with a centralized defense strategy; Forrester and his Whirlwind team found themselves trying to find a way to identify enemy bombers in the sky. Soon they developed the first visual interface for a computer. Radar coordinates, transmitted over telephone lines, were converted by software into a sequence of dots on a cathode ray tube. Next, they designed a light pen to select information from this display for more sophisticated processing. The light pen, which looked like a thick ballpoint pen with a light-sensitive diode at its tip, was connected to the computer by means of a port. Its beam interacted with the cathode rays on the computer screen to communicate information, a function that in the future was typically performed by a mouse or joystick.

More challenges had to be met. In order to create rapid access to a stable computer memory, Forrester worked to perfect an elegant, effective method for storing data, magnetic core memory. Tiny cores or rings of magnetic material were embedded in a fine latticework of connecting wires to record and communicate information. Each core could be assigned one bit of information that was easily retrieved. This memory enabled a general-purpose computer to function with unprecedented complexity and speed. The sketch for the magnetic core memory was the basic configuration to be used by computers for the next three decades, but at first the wonderful machine had no peer and no immediately clear purpose.

Meanwhile the Department of Defense hired IBM to produce a revised version of Whirlwind, the Semiautomatic Ground Environment (SAGE), as the core of a projected continental surveillance system. Stimulated by this government contract, the company had a leg up over potential competitors in the industry.

For all the complex technology required, SAGE performed a readily understood function. The computer compared all radar sightings with a file of internally stored aircraft schedules. When a blip on the screen did not match this information, SAGE calculated a trajectory for intercepting the unidentified object in flight.

The Soviet Union's test of a hydrogen bomb in 1953 exacerbated fears of vulnerability to nuclear attack, but SAGE was not ready for operation until five years later. Eventually twenty-six devices were installed at military bases around the nation. Larger than any other computers built before or since, each weighed 250 tons and used fifty thousand vacuum tubes. SAGE was manned around the clock; its operators were trained in elaborately planned simulations of enemy attack. They were in effect professional computer programmers. SAGE also sparked the invention of the modem, the de-

To track the movements of aircraft, an airman uses a light pen to interface with one of the first computer screens, which were developed for the computerized air defense system called SAGE. On the screen, the path of a plane is indicated by dots of light. The light pen reads its position and calculates a path for interception. A diode at the tip of the light pen is connected to the computer via a port, much like a mouse on a personal computer today.

vice used to convert information and send it through telephone lines from one computer to another. Few of the millions of people using modems in their homes and offices as a matter of course at the end of the century realized that E-mail from a kid in college was the legacy of Cold War anxieties about air defense.

But SAGE's strategic potential was never finally put to the test. For all its technological complexity and its huge drain upon the defense budget, it was about to be eclipsed as fast-track technological challenge by history's most gargantuan web of interconnected government, business, and academic activity.

Race into Space

Yet again the Soviet Union aroused fear and motivated U.S. technological progress. On October 4, 1957, *Sputnik 1* soared into orbit, becoming the world's first artificial satellite. Previously Soviet space technology had been playing catch-up. Had it leaped ahead of the United States at last?

Now the threat was not nuclear bombs delivered by droning bombers detectable by SAGE but swift, sleek rockets with nuclear warheads. Four weeks later *Sputnik 2*, carrying an eleven-hundred-pound payload and a live dog, was circling the globe.

Senator Lyndon Johnson's reaction summed up the national mood in America: "It took the Soviets four years to develop the atomic bomb. It took them two years to develop the hydrogen bomb. And now, there is something strange in the heavens."

The Soviet leader, Nikita Khrushchev, impishly fed U.S. concern by claiming that the USSR had "all the rockets it needs of all ranges," including an intercontinental ballistic missile (ICBM).

This next phase of feverish Cold War competition had had its origins back in March 1945, when victorious Soviet troops captured the Germans' secret rocket facility at Peenemünde and took some of the scientists prisoner. Fifty-five rocket engineers, including the center's brilliant director, Wernher von Braun, had already fled. They surrendered to the Americans, bringing along one hundred V-2 rockets.

These frightening weapons had progressed rapidly during the worldwide conflict. First launched successfully in 1942, the V-2, a missile forty-six feet long and weighing twenty-seven thousand pounds, could fly at 3,000 mph for up to 120 miles. It rose in its murderous arc to 60 miles before sliding down to target. It was the direct ancestor of all important rockets developed and flown since. No nation but Germany had anything like it, and about five thousand had been produced by war's end. Even before then

At the White Sands Proving Grounds in New Mexico, set up for U.S. rocket research just the year before, a V-2 rocket is shown being readied for firing tests on May 10, 1946. The launching platform is set in concrete, borrowing techniques developed earlier by German rocket engineers at Peenemünde. The rocket has been positioned upright by the Meiler Wagon (shown lying on the ground to the left), a huge steel elevating boom of German design.

Braun and his team had designed manned rockets and four-stage launchers for placing satellites in orbit around the earth.

Quietly 130 of Braun's Peenemünde researchers and other scientists had developed the Vanguard rocket in the United States, and a relieved nation awaited its well-publicized launch on December 6, 1957. Within seconds of firing, it exploded into shrapnel.

Two months later a second Vanguard took off successfully. Then the United States sent an Explorer satellite into orbit. But the world's attention was continually diverted from these catch-up exercises by Soviet surprises: ever more spectacular launches and ever heavier payloads. The media-savvy Khrushchev boasted that ICBMs were coming off the Russian assembly lines "like sausages." He crowed that his nation's space age weaponry was capable of wiping all "probable opponents from the face of the earth." By 1960 the Central Intelligence Agency (CIA) secretly estimated that there were at least 120 Soviet ICBMs in existence.

That year's victor in the U.S. presidential election, John F. Kennedy, warned ominously of a "missile gap" during his campaign. But by the time he took his oath of office in January 1961, the gap was revealed as myth or misapprehension: U.S. satellites first launched in the top secret Corona project the previous August now relayed images of virtually every corner of the USSR. Surprised and delighted Pentagon analysts discov-

In the search for a replacement for the relatively large, unreliable, and continually overheating vacuum tube, William Bradford Shockley discovered in 1947 that crystals of germanium could be used as both amplifiers and rectifiers, like triodes. His research team's first device, the point contact type of transistor, had two wires connected to the crystal and sealed within a case, as shown in this cutaway photo.

ered that only 4 ICBMs had been put together in Russia. The United States had stock-piled 224. In addition, 1,000 Minutemen were funded in the new U.S. defense budget.

The Minuteman—a three-stage solid-fuel ballistic missile capable of carrying a nuclear warhead from many different sites in the United States to important targets in the USSR—was fully operational within the Strategic Air Command's inventory by October 1962. Because its solid fuel was fairly stable chemically, the missile could be fueled, placed at rest in a silo underground, then fired on a moment's notice. By 1965 some eight hundred Minutemen were lying in wait near five Air Force bases.

Drawing upon the technology that led to the giant SAGE, a miniaturized computer—necessarily small and light despite its fifteen thousand components—was on board to guide the missile. Computer reliability and size were improved one hundred times by the replacement of vacuum tubes with transistors, arguably the most important advance in electronics since World War II. Developed in 1948 and made from crystals of silicon or germanium, transistors were much smaller than the clunky tubes, thus saving space, and did not have to be heated to work as diodes, thus saving energy. They are known as "solid-state" because they are manufactured from solid material.

There were no lulls in the continuing arms race. The virtue of the Minuteman was "massive retaliation," with scores of destructive warheads hitting preselected enemy targets in cataclysmic waves. When military planners designed a policy of flexible response, missiles needed computers that could be reprogrammed an instant before launch. Working together, the familiar trio of government, academia, and industry strived to perfect the integrated circuit, which eventually made flexible response a practical reality should doomsday ever arrive. The revolutionary integrated circuit could incorporate the functions of hundreds of transistors and other electronic components in a crystalline silicon chip. It was incredibly tiny—usually 0.01 inch thick and perhaps 0.5 inch square—but its real contributions were greater dependability and better performance, not to mention decreased cost of manufacture.

On the Brink

Still, the United States was considered the laggard in the race for dominance in the upper atmosphere. Few believed that the space race was pure science right now. This was a life-or-death contest between determined adversaries. No one was surprised that Khrushchev celebrated the first orbiting human, the Soviet Yuri Gagarin, by exulting in 1961, "Let the capitalist countries catch up with us now!"

Eager to reverse the world's perception and inspire his own citizenry as well, Kennedy asked his science advisers how the United States could most dramatically outdo the Soviets. Their answer built upon American strengths—that is, our unsurpassed guidance systems and highly miniaturized computers. Possibly these could lead to a huge technological leap to the far side of known Soviet potential.

Tne new President rallied the nation with a stirring call to support space technology as strongly as more familiar areas of scientific discovery and technological development: "The United States no longer carries the same image of a vital society, on the move, with its brightest days ahead as it carried a decade or two decades ago. . . . We're first in other areas of science, but in space, which is the new science, we're not first." Kennedy committed the nation to land a man on the moon before the end of the decade and to return him safely to earth. Congress was skeptical about this idea, not to mention its forty-billion-dollar price tag, and there were those who groused about technological anticommunism. But the American people were strongly supportive, whatever their exact mixture of patriotism and fear, scientific curiosity and romantic fantasy.

The Apollo program went into high gear at NASA, but the other side continued to pull off one public relations and technological coup after another: first woman in space, first two-man capsule, first three-man capsule, first space walk, first rocket to land on the moon.

The years passed. Eventually NASA engineers made an important decision to break down the moon attempt into two stages: First, the main spacecraft would fly to the ancient rocky satellite and go into orbit, and then only a small module would land on the surface. Four on-board computers, along with a complex skein of computer systems on earth, would be necessary to accomplish these maneuvers.

The summer of 1968 was not an optimistic time at NASA headquarters in Houston. The lunar landing module was scheduled to be tested in earth orbit in the fall, but technical problems insistently sprang up. The Saturn 5 rocket, developed under Wernher von Braun's supervision to make the trip to the moon, had not yet launched a

On June 16, 1963, as the United States was struggling to duplicate the spectacular public relations successes of the Soviet Union's space program, Valentina Tereshkova, a Russian cosmonaut, became the first woman in space. In a period of about seventy-eight hours, she completed forty-eight orbits of the earth. Just a month before, U.S. astronaut L. Gordon Cooper had orbited for thirty-four hours in the Faith 7, *circling the globe twenty times.*

manned mission. No American had yet traveled as far as a thousand miles above the surface of the planet, let alone the half million miles necessary to make a round trip to the moon.

Under this pressure, Apollo's chief engineer, George Lowe, had an inspiration that might advance the team's technological knowledge. Setting aside the problems of the lunar landing module for the moment, NASA could send the manned Saturn 5 to the moon in December, put it into lunar orbit, test its ability to communicate the 250,000 miles back to earth, and also test its reentry at hypersonic speeds to the earth's atmosphere. NASA Director James Webb thought the proposal very risky, but CIA reports that the Soviets were nearly ready to send their Soyuz spacecraft into circumlunar orbit clinched the decision: *Apollo 8*'s six-day mission was announced to begin on December 21.

The three crew members—commander Frank Borman, James Lovell, and William Anders—considered the Saturn's on-board computer their fourth crew mem-

ber, since it could precisely fire the rocket engine, keep its antenna trained on the earth, and align its navigation platform with the stars. Indeed the device could almost fly the spacecraft by itself, if nothing went wrong.

It made a dramatic contrast when aviation's legendary pioneer Charles Lindbergh and his wife, Anne, lunched with the *Apollo 8* crew the day before the launch. To calculate the fuel necessary to fly alone across the Atlantic forty-one years before, Lindbergh recalled, he had measured the distance from New York to Paris on a library globe with a piece of string.

Visiting the Moon

The 363-foot-high Saturn 5, taller than any man-made structure in the world at the beginning of our century, was powered by a first-stage booster rocket with the energy of an exploding atomic bomb. At liftoff it had 7.5 million pounds of thrust. Although the craft wobbled shakily into the air, with spasmodic jerks caused by its continually self-correcting engines, it reached supersonic speeds after forty seconds and glided smoothly up toward space. After eleven and a half minutes of flight the second-stage rocket cut off

and fell away along with the thruster casing; the command module slid into orbit around the earth.

As the world held its breath, NASA gave the command for translunar injection. With a long, gentle push the third-stage rocket came to life, pushing *Apollo 8* on its way to humankind's first rendezvous with the moon. The trip would describe a gigantic figure eight, with a large loop around the earth and a smaller loop around the moon in lunar orbit.

Except for Borman's bouts of vomiting and diarrhea, perhaps caused by the weightlessness of outer space, the three-day quarter-of-a-million-mile journey was uneventful. When the astronauts arrived at their destination, a mere four minutes of rocket fire set the craft in orbit around the moon just sixty-nine miles above its surface. A few seconds of extra firing would have slowed the ship too much, letting it slam down to the lunar surface. The computer was programmed to fire accurately, but Borman took control, manually pushing the shutoff button himself. The spacecraft coasted toward the far side of the moon. At NASA there were gallows humor jokes about the seventy-mile-high mountain there. But the other side of the satellite, never before seen, much less probed, by beings from the earth, offered no surprises.

Somewhat disappointed by the gray, rocky expanse of the 2,160-mile-wide moon, particularly by the "dark side" that is as seamlessly desolate as the rest, Anders said he had traveled through space to see little but "dirty beach sand."

But on the fourth orbit, as Borman maneuvered the ship, the crew members gasped in unison at an unexpected sight. Infinitely alone in the vast ink black sky, a radiant half circle of blue and white rolled slowly into view.

Almost immediately the image of the earth as a fragile jewel pendant in vast space became a powerful catalyst for the environmentalist movement in America and elsewhere.

Less than a year later two American astronauts, Neil A. Armstrong and Edwin E. "Buzz" Aldrin, became the first humans to walk on the surface of the moon, while six

On the Apollo 8 *lunar orbit mission launched on December 21, 1968, James Lovell piloted the tiny command module. He and his fellow crew members, mission commander Frank Borman and William Anders, orbited the earth's lone satellite ten times. They were the first human beings to see the far side of the moon, which turned out to hold no surprises.*

In a photo taken from the Apollo 11 *spacecraft in 1969, the earth rises above an area of the moon known as Smyth's Sea. This was the flight during which a human being first walked on lunar terrain. After landing in the lunar module, which became world-famous as the* Eagle, *Neil Armstrong was first to stand on the surface; then Buzz Aldrin followed. Above them, Michael Collins orbited in the* Columbia, *the command module.*

hundred million people on the blue planet watched on television sets at home and in public gatherings.

In *A Man on the Moon: The Voyages of the Apollo Astronauts,* Andrew Chaikin recalls the significance of this telecast: "It was something that had never happened before in the entire history of exploration, because for the first time millions of people were watching and listening as it happened. That little TV camera had transformed the experience. Suddenly, this trip to the moon became something that people could actually share sitting in their livingrooms, and it was Christmas Eve."

Then, anticlimactically, the space race that had consumed such vast resources, stimulated so many technological strides forward, and fed such depths of national pride and fear began to fade into textbooks and video archives. The world's attention turned back upon itself, and it was the computer, so silently and rapidly keeping space exploration on course, that became the supreme technological influence of the last decades of the century.

Toward the Twenty-first Century

As the world worked at redefining itself after the "collapse of communism," technology was powerfully shaped by the new economic, social, and political realities. Previously, the tensions of the Cold War had propelled techological developments, usually in the direction of real or perceived military requirements and often in great secrecy. In a world with only one superpower left upright, however, consumer needs and wants became the engine that now drives technological innovation.

Only one nation could afford to consider missions to Mars and beyond, and the people of that nation—worried in part about the decay and obsolescence of the infrastructures that supported the technologies of land and air transportation, in part about a national debt described as astronomical, and not at all about the possibility of an enemy's settling on other planets to do ill—no longer enthusiastically agreed that their tax dollars should be launched into space. Now it seemed that the technological advances of the next century were more likely to occur right here on earth: in our offices, classrooms, hospitals, airports, and homes.

The blossoming Internet—useful to many, almost addictive to some, frightening to a suspicious few—began very modestly. In 1969 the ARPAnet, designed as a cooperative network of time-sharing computers set up by the Advanced Research Projects Agency (ARPA) of the U.S. Defense Department, linked computers at four university campuses. The machines were of different types, but ARPAnet enabled them to communicate with each other. Four years later the system was connected to computers in Western Europe. Eventually, this ground network was connected to radio and satellite networks, creating a network of networks.

But the scientists given access to ARPAnet were changing its profile every time they went on-line, for it was the E-mail function they used most frequently. The sophisticated government-supported communications network became a high-speed digital post office for all kinds of information sharing, news, gossip, and playfulness. By 1976 even Queen Elizabeth II of Great Britain was posting E-mail.

ARPAnet was left behind and is used today in government research. Its progeny covered the globe. Toward the end of the 1980s the market for super-minicomputer industries and the home personal computer expanded greatly. The so-called Internet, based on a common language for all participating computers and featuring newsgroups in which self-selected participants could discuss a particular topic, was soon being used by individuals, institutions, and corporations. CERN in 1991 inaugurated the World Wide Web, a revolution in communications that allowed host computer programmers to set up Web sites that eventually offered photos and sound as well as words to anyone who "hit" the sites. Companies, fringe groups, political candidates, fan clubs, magazines, churches, and thousands of creative individuals set up home pages. By 1996 the so-called information superhighway included almost ten million computer hosts in perhaps 150 countries, and the tentacles of the Web continue to spread. At certain times of the day the wires become so clogged with eager networking that instant communication becomes a waiting game.

Serious questions are raised by this stunning technological phenomenon, which almost seems to have generated spontaneously, leaving the human imagination to catch up. On the one hand, any schoolchild can quickly access information on virtually any subject in this worldwide electronic library, but material generally thought unsuitable for nonadults is just as readily available. On the one hand, hobbyists can share information and oppressed peoples can publicize their grievances worldwide, but ugly rumors and crazed conspiracy theories can poison the body politic with a thousand points of hate. Moreover, while middle-income households in developed countries are thrilled by the instant communication with people around the globe, the great bulk of humankind cannot share in the fun and may never be able to do so. Will an even greater gap yawn between the informed and the uninformed, a communications deficiency eventually more dangerous than income disparities?

Unlike radio and television broadcasting, the Internet and World Wide Web have not yet found an identity that yields profits, although some goods are sold directly over the Net, occasional stunts like a "live" Rolling Stones concert gain attention, and most large corporations maintain Web sites, if only to ensure that they're on the scene when profitability strikes.

"The future of the Internet will probably be a new kind of entertainment," says Doug Rice, president of Manhattan-based Interactive 8, a company that produces corporate Web pages. "But no one knows what it will be. Because users self-select, they don't always watch what a company wants them to watch. The real story may lie with virtual communities, since there are now twelve million or so newsgroups, but that raises another question. . . . Will people withdraw from the real community of family and relationships to chat their lives away on the net?"

If the previous ten decades are any guide, the Net will evolve unpredictably, as the needs and whims of human beings intersect with such expected technological developments as switches that can handle ten times as much data, Internet telephones, and so-called smart terminals that will connect personal computers to extremely powerful host computers.

From Cold War concerns at the Pentagon to university labs where researchers redefined a communications network to a dizzying flood of new communications uses and emerging technologies, the Internet is the future in a form that is familiar to us now, an idea that grows through technological innovation into a system that can change the world of human experience.

CHAPTER THREE

*Lava burning out of the ground
at Volcanoes National Park on
Hawaii's Big Island is part of the
ancient process of forming huge
mountains from ovens deep within
the earth and beneath the sea.*

ORIGINS

MORE ANCIENT, LESS STABLE

About one hundred years ago many people still believed that the physical earth was created in seven days of twenty-four hours each and that the extent of the seven seas, the heights of mountains, and the depths of valleys had been set in stone for all time. According to the calculations of Bible scholars in the Middle Ages, who relied chiefly upon the generations and dynasties chronicled in the Old Testament, this extraordinary week occurred some four thousand years before the time of Jesus.

Human beings—first Adam, then Eve—were created during that week, as were all the plants and animals around the globe and beneath the seas. The answers of religion combined "common sense," it seemed, with the occasional incident of divine intervention. The answers of religion were apparently comprehensible to and comfortable for most people in the West, but scientists were not satisfied.

At first slowly, then with confusing speed, they began to gather clues that contradicted this straightforward picture on many levels. The age of the earth eventually was proved to be almost inconceivably ancient; the beginnings of humanity were discovered far back in time; and the stable visible facts of the earth—its physical characteristics, its flora and fauna—were revealed as fluid and constantly changing.

"Millimeters or Centimeters"

By the end of the nineteenth century European scientists who specialized in geology— the science that focuses on the history, structure, and physical makeup of the earth— were convinced that all previous estimates of the world's age were way off.

All over the world they had been able to measure layers of rock, known as strata, that were clearly visible in river gorges or other areas of exposed rock. They knew that such strata were produced by the gradual accumulation of earth, sand, and pebbles washed downstream by rivers and deposited on riverbeds or on the ocean

floor. To look at the different ribbons of rock displayed in the walls of a canyon was to see the history of the formation of the world.

But careful records of the pace of this process, known as sedimentation, produced an unexpected conclusion: The buildup of the earth's rock layers would have taken much longer than anyone had ever thought possible. Professor Ralph Harvey has explained how this astonishing discovery arose from the humblest kind of scientific observation: "You could go out and measure the sedimentation rate—that is, the amount of sediment typically carried by a river and the accumulating thickness of sediment on the river bed—and what you found was only millimeters or centimeters at the most during a normal human time-span."

This observation encouraged researchers to travel ever more widely in search of data. What they found affirmed and strengthened the basic line of reasoning: In some areas the layers of rock could be measured in kilometers, but everyone knew how infinitely slowly the floor of the sea at a favorite summer resort or the bed of the river flowing past a college campus was being built up by sediment.

Nineteenth-century scientists recognized that the geological record was indisputable: The earth began forming millions and millions of years before their tentative measurements and surprised conclusions. Some estimated that the age of the land beneath our feet must be between three to five hundred million years.

These estimates were calculated with more precision in our own century.

Only Ninety-eight Million Years Old

Geologists became comfortable with the concept of a globe at least hundreds of millions years old after a while, but their shared estimates suffered a setback when Britain's Lord Kelvin, an undeniably brilliant physicist, "proved" that the earth was *only* ninety-eight million years old.

Even on the odyssey of science the greatest of thinkers can occasionally be blind to upsetting data or the types of analysis used in a different scientific discipline. In a long and distinguished career Lord Kelvin had never noticeably been wrong. But to date

The beautifully striated canyon walls above the Fremont River in Utah's Capitol Reef National Park reveal sedimentation over aeons, a natural gauge used to help measure the earth's vast age. The concept was discovered and explained in 1086 by Chinese scientist Shen Kua, but Western geologists would not fully recognize the stunning implications for earth history until the nineteenth century.

the earth as he did, he had to ignore a century of work by geologists. Perhaps arrogance was only one factor. In the nineteenth century it was still possible to believe that an educated human being could know almost everything there was to know; in the twentieth century science is so specialized that areas of expertise grow paradoxically more complex and more restricted at the same time. Lord Kelvin was able to play the role of generalist.

Born William Thomson in Ireland, the future knight of British science was recognized as a math prodigy in early childhood. At age ten, he was accepted into the University of Glasgow, and he published his first influential scientific paper six years later. At the unusually young age of twenty-three, he was named a professor of natural philosophy at the university, where he was to teach for more than half a century.

Like the other Newtonian physicists of his day, he concentrated on deepening science's understanding of heat, energy, and the other physical measurements of the world and made many contributions to the science of thermodynamics, the study of heat and how it converts to energy. He developed the Kelvin temperature scale, which measures heat in terms of absolute numbers—for example, 273 degrees for the freezing point of water, 717 degrees for the boiling point of sulfur.

Left: *Lord Kelvin, a Scottish physicist who was extremely gifted as both theorist and technological innovator, bestrode British intellectual life with such supreme self-confidence (and became so wealthy from his inventions) that lesser scientists accepted his dismissal of geology, which he found as intellectually weighty as stamp collecting.*
Right: *Meteor Crater in Colorado, which exhibits the classic bowl shape seen in impact craters photographed on other planets and the moon, was formed some twenty-five thousand years ago when an extraterrestrial mass of iron slammed into a plateau of sedimentary rocks. According to one theory discussed early in the twentieth century, the earth was formed by the gradual accumulation of meteors and other such extraterrestrial debris.*

In addition to his gifted theorizing, this particular professor could translate his intuitions in physics into practical devices for general use. His inventions for measurements of energy and heat and for navigation earned him royalties that purchased a grand country estate in Scotland, reportedly one of the first houses in the nation to be electrified. In 1856 he became famous throughout the world as the chief scientist for the first project to lay a telegraph cable across the Atlantic from Europe to the United States. His understanding of how electricity is conducted was crucial to making the project work.

In short, Lord Kelvin had the reputation, the self-confidence, and the track record to make much of the world regard his conclusions as something like holy writ.

Why did he arrive at the comparatively youthful age for the earth? As a physicist concerned with heat-related issues he knew well that the earth becomes hotter the deeper one digs down. Miners were familiar with this phenomenon, of course, and Lord Kelvin used their measurements of increasing temperatures.

But his interpretation of these data was influenced by a wrongheaded premise: He simply believed, without proof, that the earth must be a large piece of sun that was somehow broken off and hurled into orbit. Over the millennia it would have lost heat.

Therefore he need only determine how long it would have taken to solidify and cool to its present temperatures. His unwarranted assumption, combined with his flawed interpretation of minimal data, convinced him that ninety-eight million years was the answer.

From hundreds of millions back down to this lower exact number, the theorizing ranged widely late in the nineteenth century, but Kelvin's conservative estimate became the standard among the intelligentsia. Unfortunately the dazzle factor of his remarkable career blinded too many people to the evident facts.

A geologist at Princeton University would not let himself be numbered among them. Thomas Chrowder Chamberlin knew that the fashion for lowering the earth's estimated age was mistaken, that Kelvin's assumptions were questionable and therefore his results unreliable.

Chamberlin pointed out that large numbers of meteorites had been observed in our solar system. The earth could have been formed by the gradual accumulation of these wandering rocks, an assumption with no less to recommend it than Kelvin's assumption that a bit of hot gaseous sun was ejected into space and became our solid-seeming planet.

Even the unaided human eye can detect meteor craters on the airless moon; the telescope finds thousands. In the eons of the earth's existence, it was possible that innumerable meteors and asteroids had struck our home planet. Perhaps the physics of the collisions and the buildup of mass explained the rise in temperature beneath the crust.

Nothing was proved by Chamberlin's objections, as he knew, but he believed that he brought the discussion back to reality. No one yet had concrete evidence to show how and when the earth was formed. His theory was as theoretically sound and evidence-free as Lord Kelvin's. In short, science could not yet use scientific reasoning to reach general agreement among the disciplines about the age of the earth.

Strange Rays

Then the world began to learn about some very strange materials. A sensation occurred just before the beginning of the twentieth century, when Wilhelm Conrad Röntgen, a

The very short waves known today as X rays, a form of electromagnetic radiation, were discovered by Wilhelm Röntgen, in an experiment with fluorescence. Nonscientists were fascinated by the penetrative luminescence revealed in this X-ray photo of his wife's hand. The discovery would revolutionize the practice of medicine and earn Röntgen the Nobel Prize for Physics in 1901.

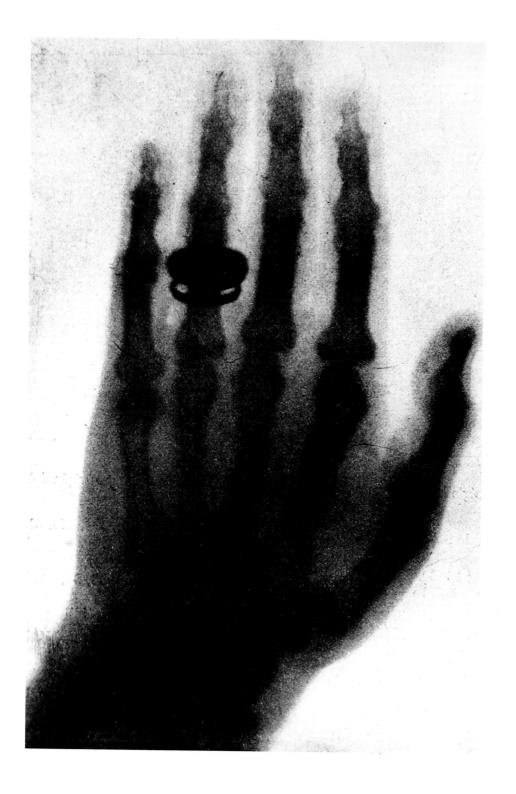

German physicist, accidentally discovered a shortwave ray known today as the X ray.

In 1895, he was trying different pieces of lead in the glow of a cathode-ray tube when he suddenly froze in wonder. In the darkness of the laboratory the outline of his thumb and forefinger was irradiated with dim cold light, and within the outline, for the first time in history, a human being saw the bones within the skin.

Röntgen figured out that the invisible, highly penetrating rays emitted by the anode must be some kind of electromagnetic radiation with a much shorter frequency than visible light. Yet because a magnetic field had no effect upon these rays in experiments, they must resemble light waves instead of electrically charged particles.

Such images as the bones in Röntgen's hands were both amazing and a little frightening to the public. It was reported that seaside vacationers began swimming fully clothed, for example, in order to prevent prurient scientists from somehow photographing them in the nude.

Unstable Nuclei

The rage for X rays stimulated various avenues of new research, but none more fruitful than the experiments of Antoine Henri Becquerel. Like many other scientists in the nineteenth century, he was fascinated by luminescence, or any glow produced in nature without heat. Perhaps the eerie X rays would provide a clue. In his Paris laboratory Becquerel looked for traces of the rays in luminescent crystals. Perhaps these naturally

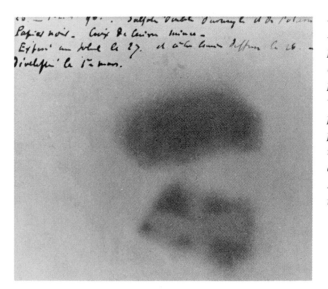

Inscribed in French with Antoine Henri Becquerel's lab notation, this plate was one of those involved in his discovery of radioactivity. This plate and several others were wrapped together with deposits of a phosphorescent uranium salt and put aside in a dark drawer; this image, intensely burned into the plate, appeared when it was developed. Becquerel's lucky breakthrough led to the use of radioactive dating.

glowing minerals, he reasoned, were spontaneously emitting the same or similar rays as the anode in a cathode-ray tube.

For some time he seemed to be getting nowhere, although his method was ingenious enough. He sprinkled a variety of the luminescent crystals on photographic negatives wrapped in opaque paper to protect them from light, then exposed the negatives to sunlight and developed them in his darkroom, looking for the mysterious rays. There was no sign of progress.

But chance favors the bullheaded. One day, when he had placed crystals of a uranium salt, potassium uranyl sulfate, on a negative and set it outside, Paris fell victim to one of its famously dreary stretches of leaden skies. Becquerel, bored to distraction, decided to while away the murk by developing the negative. He knew that there would be no X rays, and he was right.

Instead he discovered something much more astonishing and much more significant for progress along the odyssey of science: the existence of radioactivity. The plate showed an outline of the crystals; later Becquerel found that the salts would burn an image into a negative even in a closed drawer. In other words, these odd rays were not affected by sunlight. Something was being naturally emitted by the salts, and it was strong enough to pierce through paper or wood to affect the emulsion on a photographic plate.

By 1900 Becquerel understood that the element of uranium alone was producing the phenomenon. Moreover, more than one type of radiation was involved. He isolated beta particles, which are high-speed, highly penetrating electrons or (rarely) positrons, and gamma rays, a kind of electromagnetic radiation with such penetrative capability that three feet of concrete or up to ten inches of lead are required to deflect it. The third component of radiation, ordinary helium nuclei known as alpha particles, was not discovered until later.

In short, radiation is the spontaneous emission of energy in the form of these waves, which beam out from rocks found in the earth. Although Becquerel did not realize it, uranium is a radioactive element that characteristically decays at a steady rate. Such elements decay into lighter "daughter" elements, releasing their energy in the form of radiation.

Ticking Rocks

The study of radioactivity intrigued many physicists, but it was the legendary Polish-born scientist Marie Curie and her French husband, Pierre, who applied the concept to the continuing puzzle of the earth's age.

Madame Curie saw that radioactivity, a term she coined, was the emission of particles when the nuclei of atoms decay. She learned to refine radium, which is four times as radioactive as uranium oxide, from a naturally occurring substance called pitchblende. The Curies discovered both radium and polonium in this mineral, but radium gave off more radioactivity in its eerie greenish glow. Vast quantities of pitchblende yielded only minute amounts of radium after being refined and reduced chemically. It took Madame Curie a decade, but by 1910 she had isolated pure grains of the element.

Radium, it turned out, was a kind of radioactive clock, a rock that began ticking as a measurement of time in the geological record from the moment it was formed. As radium or any other radioactive element ages, it loses its radioactive strength. Each element has a signature half-life, the number of years required for it to lose half its radioactivity. The halving time of different elements varies hugely, from millions of years down to fractions of a second. Therefore, it is possible to figure out the age of a rock or mineral by radiometric dating: Scientists measure the amount of radioactive element remaining, compare it with that element's known constant rate of decay, and the ratio established when the rock or mineral was formed.

The radioactive clocks proved Kelvin wrong and nineteenth-century biologists too cautious. The average age of the earth's oldest rocks seemed to be about two and a half *billion* years. This was a calculation based upon fact, not assumption. However the world was formed, the event had happened so long ago that human history on the planet suddenly seemed no more than the flick of an eyelash.

Much, Much Older

But the earth was older yet.

Arthur Holmes, who was to become a professor of geology at the University of Edinburgh, started his breakthrough research in geology in 1911, when he was a physicist at London's Imperial College. Previously, radiation dating was comparatively primitive, but Holmes became the first scientist to realize that the technique could be enormously important in dating rock with greater precision. In 1913, when he was only twenty-three years old, he published his first sophisticated research, a book titled *The Age of the Earth*. Holmes became the first person to recognize—after he mastered all of the studies thus far made of radiation—that the uranium discovered in the previous century by Becquerel would be the most reliable possible radioactive clock. As uranium decays, eventually turning to lead, it leaves a highly accurate ratio of uranium to lead in a rock sample.

Shown in 1925 working at the Radium Institute in Paris with her daughter Irène Joliot-Curie, the Polish-born Marie Curie and her husband, Pierre, discovered radium and polonium. She was also involved in proving that beta rays are made of high-speed electrons. In 1906, after her husband's death in a traffic accident, Madame Curie was appointed the first woman professor at the Sorbonne.

By the 1920s Holmes showed that earth was vastly older than studies of sediment had yet suggested, something like 4,400,000,000 years old. In about a quarter of a century, in other words, science's agreement about the age of earth had increased by a multiple of at least one thousand. This was an explosion of knowledge that was also an implosion of ideology. Humankind and its home were not at all what we had thought and held dear for centuries.

The Dance of Continents

Had the earth sat, unchanging, for those billions of years? Very little was known at the beginning of the twentieth century about the history of the surface of the earth. The mountains and seas and rivers mentioned in the Bible were still there, apparently unaltered after millennia. No civilizations in other parts of the world had left records to suggest that a mountain range had worn down or a great sea dried up.

As for the regions beneath the surface, much was guesswork. Scientists agreed that a rigid, fixed crust covered the earth, visible as the dry land of continents and islands but mostly acting as the unseen floor of the oceans that blanketed 70 percent of the globe. Below that thin crust, in their view, there lay a thick layer of mantle that they surmised to be composed of hot, fluidlike molten rock and magma. Finally, they believed that there was a kind of core of some sort at the center of the earth.

But what exactly was the nature of the crust, the mantle, the core? There were so many fragmentary notions that geological science had room for five different but equally plausible hypotheses about how mountains formed.

Yet it was a meteorologist making weather observations, not a geologist, who uncovered one of the most important mechanisms of the formation of the earth we see today. Early in the twentieth century the German weather expert Alfred Wegener joined an expedition to Greenland to study climate in the little-known Arctic region.

Like many other scientists who have changed history, Wegener did not slavishly restrict himself to observations in his own field. As he knew, mapmakers had been intrigued for centuries by the way the coastlines of Africa and South America looked as if they could easily fit together. Still, no one thought that the two great continents had ever actually been joined.

Wegener took the leap, writing an article that made just that argument in 1912. Three years later he published the first of his many books on the subject. His evidence included an informed look at the geological record in relation to climatic conditions.

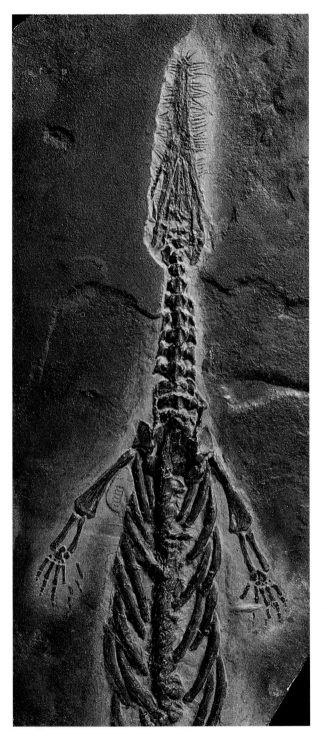

The front end of a fossil skeleton of the tiny Mesosaurus *from the Karoo Desert in South Africa provided an important clue to meteorologist Alfred Wegener, who ranged out of his field to stun academic geologists early in the twentieth century. How, he wondered, could this tiny creature have existed in prehistoric times in both Africa and South America? If the creature was too fragile to travel that far, he reasoned, then perhaps the continents traveled.*

For example, great wide striated pavements of rock had been found in the Karoo Desert in South Africa, but such formations were typically produced by the scoring, smoothing action of glaciers moving slowly across them. As Sir Ronald Oxburgh has explained, "Everyone knows that you can have glaciers in tropical areas if you've got very high alpine-type mountains. But great, extensive sheets of ice can't exist there, yet that's what the geological evidence seems to show."

That wasn't all. Geologists in Wegener's day knew that coal was most likely formed in tropical climates some three hundred million years before. What, then, were large coal deposits doing in the frozen islands of Spitsbergen, which lie far above the Arctic Circle?

In addition, there was the odd fossil record of *Mesosaurus,* a small aquatic reptile found only in Brazil and southern Africa. Could this weak prehistoric creature have swum three thousand miles across the Atlantic? Was it merely coincidence that the rock formations in the two separate fossil sites were similar in structure and both three hundred million years old? Wegener thought not.

Wegener became firmly convinced that he now had hard empiric evidence that provided at least half the answer to these mysteries: his theory of floating continents. Two or three hundred million years in the past, he argued, all the continents joined together in one intact, very large supercontinent, which he named Pangaea. It split into two major landmasses, Laurasia and Gondwanaland. Then the continents known to recorded history were formed at about the end of the Mesozoic Era, sixty-five million years ago. The American continent slid westward away from the European landmass, and the Atlantic poured in to fill the gap. India and Australia drifted off from the eastern shores of Africa. Moreover, Wegener speculated that mountain ridges were formed by friction when the leading edges of the moving continents met with each other.

The other half of the puzzle he could not answer well: Why did the continents drift ever so slowly apart, making a new outline of the continents and seas? He suggested that the earth's rotation somehow affected a flexible layer beneath the crust, causing the drift. He was on less solid ground in theorizing this way about causation.

The Missing Force

Although Wegener continued to reargue his theory in detail in book after book, pointing to new bits of geological evidence to support his conclusion, the scientific community did not take him seriously. We may forget that any age produces—and publishes—crackpot theories to explain the most puzzling mysteries. Sometimes the nuttiest books become best sellers, enriching their imaginative authors, then fade away.

Geologists put Wegener in the crackpot category for understandable reasons. In the first place, his best evidence for floating or drifting continents lay along the scarcely populated southern shores of Africa and South America. The Western European and North American geologists who dismissed him had for the most part never seen these areas and the dramatic physical correspondences between them.

Second, geophysicists knew that the crust was a solid continuous mass all the way down to its boundary with the mantle. In addition, they knew that the continental crust was thicker than the oceanic crust and would have to plow through the latter obstacle in order to "drift" in the way Wegener theorized. But there was no mechanism known that could power such titanic thrusting.

Third, there was the conventional portrait of the mechanism of earthquakes. Scientists knew that a major tremor causes the earth to ring like a bell. For hours after-

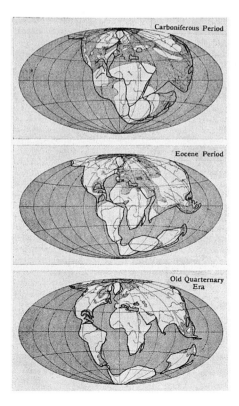

In 1922, Wegener published his theory with this illustration suggesting how one supercontinent, which he called Pangaea, had split into huge chunks of land that drifted farther apart to form the configuration of continents we know today. The reaction of established science ranged from ridicule to outraged rebuttal. This visionary theory contradicted some seven decades of accepted theoretical progress in geology.

ward long waves from the earthquake travel completely around the earth's surface. But slabs of land do not go sliding away toward another continent, even under such tremendous force. In other words, whatever force holds the earth steady during and after a powerful earthquake, they reasoned, had to be strong enough to keep the continents from drifting about. As historian of science William Glen has explained: "The soft underbelly of Wegener's 1915 theory was the fact that he did not have a force by which to split Pangaea and then propel its pieces across the face of the continent, across the face of the globe to take up their positions as the modern continents."

Finally, the scientific community was conservative and comfortable on the one hand, yet also proud of its revolutionary discoveries. It had not been so many years ago that it had been forced to defend its own radical ideas. The geologic establishment decided that traditional explanations could be found for any puzzling distribution of fossil creatures and distinctive rock strata throughout the world, even if special pleading was necessary to cover the theoretical fault lines.

In 1926, Wegener's latest evidence and arguments were presented at a meeting of the American Association of Petroleum Geologists in New York, but he was bitterly attacked, even personally ridiculed. After all, the eminent academics of British and American geological science had spent entire careers publishing academic articles based upon the assumption that the continents were fixed. Who was a weatherman (and not least, in the aftermath of World War I, a German) to undermine the received ideas of the academy? One participant's dismissive reaction was typical: "His method is not scientific, but a selective search through the literature that ignores most of the facts that are opposed to the idea." Neither the high priests of traditional geology nor Wegener could know that this debate would continue for the next fifty years or so.

Meanwhile Wegener was still a working meteorologist. In 1929 he was asked to lead a new scientific expedition, his fourth, to snowy, misnamed Greenland. The plan was to establish a base station on the east coast of the huge island, then travel on specially developed motorized sleds to the heart of Greenland's great land glacier. There, at

As courageous in action as he was in original thinking, Wegener (left) decided to dare a dangerous trek across icy Greenland in 1930, accompanied by his Inuit guide and companion Rasmus Villumsen, to deliver lifesaving supplies to researchers at a field station unexpectedly isolated by an early winter blizzard. The two men died in the attempt. Newspaper obituaries praised Wegener's heroism; by the middle of the century, he became much better known as the hero of the contemporary revolution in science known as plate tectonics.

a new station on the central icecap, two men were to conduct research into weather patterns throughout the long Arctic winter. On this expedition Wegener proved as courageous in action as in theoretical debate. Visiting the glacier research station, he was surprised by unusually early blizzards. Since there was only enough food for the two men assigned there, he volunteered on November 1, 1930, his fiftieth birthday, to return to the coastal base station, accompanied by his Inuit companion Rasmus Villumsen. They never reached the base. In the spring Wegener's body was found in a grave evidently dug by Villumsen, whose body was never recovered.

Wegener was honored as an authentic hero, but his intellectually daring theory of floating continents was set aside by the academy.

There was one important exception. Arthur Holmes, the geologist who had been instrumental in establishing the earth's great age, had been listening to Wegener with an unusually open mind. He had some complementary ideas of his own.

By the 1920s he was speculating that the unseen, unknowable interior of the earth was capable of behaving like a fluid. Making guesses that proved remarkably accurate decades afterward, he published a paper with illustrations of possible convection currents inside the earth. Such currents occur when a liquid or gas heats up. A warmed or warming fluid expands—thus becoming less dense—and then begins to rise as a current within the surrounding cooler fluid. Holmes was so certain of his concept that he charted where and how convection currents were likely to act in the liquid core of earth. One immediate implication: The structure of the earth was not as fixed as traditional scientists believed.

Still, Holmes or no one else had yet found a force powerful enough to dislodge and move the continents that merely vibrated to the waves caused by earthquakes.

"No Smoking"

Anecdotes about personality survive less because they're true than because they embody a truth. Supposedly the quietly humorous Harry Hess had taken a job as a lecturer in geology at Harvard College. On the day he was to hold his first class he discovered that

Geologist Harry Hess learned by using echo sounders on his troop transport ship in World War II that the bottom of the sea was definitely not, as conventional science assumed, a featureless plain. In 1962 his paper "History of Ocean Basins" argued that the continents had once been one, and the world knew at last that Wegener had been right.

the lecture hall displayed a huge No Smoking notice. A chain smoker, he turned on his heel and took a job at Princeton, where gentlemen were allowed to pollute the air however they chose.

Hess did as much as anyone else in the twentieth century to unlock the secrets of the earth hidden beneath the oceans. In the 1930s he became involved in researching the land beneath the seas with F. Andries Vening Meinesz, an eminent Dutch geophysicist who was firmly convinced that the notion of drifting continents was "very unlikely." As was not unusual in science, the skeptic accidentally led the way to prove his skepticism misguided.

On their first expedition Hess and Vening Meinesz attempted to obtain a clear picture of the shape of the earth beneath the southwestern Pacific by measuring variations in gravitational force. Later they boarded a U.S. submarine to study the seafloor in the Caribbean region.

Only in the submarine, beneath the rocking of the waves, could Vening Meinesz use his extremely sensitive, specially built gravimeter. It had carefully weighted pendulums that swung in the direction of the heaviest, most massive rocks in an undersea region, where the gravitational pull would be strongest. Because the variations in pull

were extremely tiny, the gravimeter was built to detect differences as small as one millionth of the earth's gravitational pull. Painstakingly assembling thousands of measurements with the gravimeter, the two scientists aimed to map large sections of the seafloor.

The results surprised them. At the Pacific margins of continents there were large areas of inexplicably weak gravity. It was as if huge undersea rock formations were being held down by some undiscovered brute power.

What could it mean? History and Hess's sudden change of career delayed the answer to that question for a few years.

The man impatient with smoking restrictions found himself seething with frustration on the submarine. As a civilian Hess had to go through the naval chain of command to get anything done. If he was seated next to the helmsman and the gravimeter's actions suggested it would be helpful to veer to starboard, he had to ask an officer to tell the enlisted man to turn the ship. In 1936 Hess was able to convince his local congressman to have him commissioned an ensign in the U.S. Navy Reserve. The day after the Japanese attack on Pearl Harbor in 1941, he was called to active duty.

After an assignment tracking German submarines in Atlantic waters, he was transferred to the Pacific, where he commanded the U.S.S. *Cape Johnson*, a marine transport and landing ship that supported invasions of Japanese-held islands. Hess was in battle many times, but he took advantage of lulls in the action to continue his scientific studies, frequently using new technology to take echo soundings of the ocean floor.

There he made his second unexpected discovery: On the ocean side of the arcuate island chains of the western Pacific were deep trenches. The vast seabed was furrowed, in other words. Moreover, these underwater valleys, which eventually were measured to a depth of seven miles below sea level, seemed to be linked with those areas of weak gravity that he and Vening Meinesz had discovered.

Still More Paradoxes

After the war large areas of the seabed were accurately surveyed and mapped for the first time. A vast, unsuspected landscape of high mountains and deep trenches was slowly revealed where, aside from the submerged mythical island of Atlantis, humans had long thought the seabed flat and featureless. In 1947, Columbia University's Maurice Ewing led an expedition to map one of these amazing geographic formations, the Mid-Atlantic Ridge with its undersea peaks of six to ten thousand feet. Now known to be part of the world's longest mountain range, this ridge connects to other ridges that circle the globe in a forty-six-thousand-mile-long system. Ewing and his colleagues discovered something very peculiar in the area: No rock found nearby was more than one hundred fifty million years old. In other words, the seafloor was very, very young in earth terms.

Back at Princeton, Hess was led by this finding to make a series of inspired guesses. Suppose the ridge marked a huge rift in the ocean crust . . . and suppose further that part of the oceanic lithosphere was rising up through this crack because it was hotter—thus, more expanded and less dense—than the cooler, older oceanic lithosphere making up the floor of ocean basins. From his studies of earthquake focal mechanisms along the mid-ocean ridges, he realized that they were caused by tensional stress: In other words, it appeared to him that the ocean crust was being pulled apart as the seafloor spread to either side. In that case, the continents were being pushed away from each other.

Moreover, if young rocks were being formed undersea near the ocean's ridges, this new material in the crust was needed to fill in space as the seafloors spread apart. Further, Hess theorized that this crust traveled underwater, then dipped back into the earth, a process called subduction, when it ran up against the edge of a thicker continental plate. The crust was forced under the plate and down into the earth's more moltenlike interior, possibly to be recycled in the mantle. This extraordinary geophysical process would explain the trenches he had discovered during the war in the Pacific. Still, Hess was cautious. Although he was attracted to this picture, he wrote about it as "an essay in geopoetry."

Concrete evidence for his speculations was already accumulating. In the Sierra Nevada mountains in California in 1960, Allan Cox and Brent Dalrymple had been chasing down an odd anomaly. The volcanic rock known as basalt was known to take on the earth's magnetic field after it flowed above the ground and cooled. But why did some basaltic outcrops have magnetic fields that pointed southward, directly opposite to the earth's magnetic north pole?

The answer was simple but at first seemed highly improbable: Every million years or so, to judge from the ages of the rock they studied, the earth's magnetism must have completely reversed. At times, the earth's magnetic pole lay in Antarctica. Eventually, it became clear that this "flip-flop," as Cox called it, has occurred nine times over the previous four million years—and at irregular intervals lasting from about one hundred thousand to a million years.

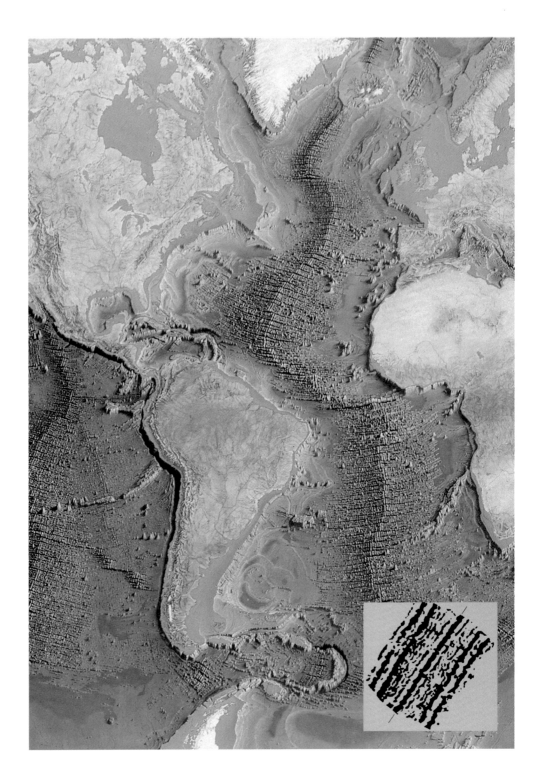

Meanwhile, Fred Vine and Drummond Matthews in Cambridge, England, were making discoveries of strange "zebra markings" beneath the sea. Magnetometers had mapped alternating stripes of magnetism that were usually several miles wide and were generally found parallel to undersea ridges. Now it was time for Vine and Matthews to make some daring surmises of their own. What if the stripes were an underwater record of the magnetic flip-flops found on land in California . . . and suppose they were parallel because they were moving away from the ridges on a kind of titanic conveyor belt?

In other words, the magnetic stripes might prove that the seafloor was spreading, just as Hess had theorized. When the two scientists published their work in 1963, they were ignored. Two years later, however, Dalrymple was able to date a magnetic reversal to nine hundred thousand years ago just as Neil Opdyke, a researcher at the Lamont Geological Observatory in Palisades, New York, discovered a magnetic stripe of the same age in magnetometer readings taken on the Pacific seafloor.

This was proof from both land and sea that the poles do flip-flop, that these reversals can be charted in the stripes on the seafloor, that the seafloor inexorably spreads.

Today the implications of these insights are known as plate tectonics, a theory that unifies all the isolated geological information that puzzled Hess and unites ocean and continental geology in one grand scheme. Scientists have been able to measure the light rock of the earth's lithosphere at an average ninety-three miles thick beneath the continents and forty-three miles beneath the seas. Constantly shifting, the seven major plates of this lithosphere move stubbornly against and under one another, causing the continents to break apart and float away over the millennia, pushing up mountains and creating new oceans. Today, the Atlantic Ocean is spreading at an estimated rate of 0.4 inch a year, with little or no subduction to offset the expansion. By contrast, the Pacific spreads faster at an annual rate of 4.3 inches, but subduction there more than compensates for this growth, so that the world's largest ocean, at about eleven thousand miles wide, is actually shrinking by the year.

This geologic ferment also explained the formation of earth's visible mountain

This detail from the dramatic undersea map created by Maurice Ewing's associates Marie Tharp and Bruce Heezen reveals the highest mountain range on the planet clearly snaking underwater between the continents. The inset diagram shows the magnetic lines lying parallel to the undersea Reykjanes Ridge, an essential piece of the puzzle that was worked out in plate tectonic theory.

ranges. For example, as the continents began to split apart four hundred million years ago, terrific pressures built up on land where continental lithosphere was colliding with oceanic lithosphere, causing the rise of the Rockies and the Andes in the Americas. The Himalayas were produced when the Eurasian plate collided with the Indian plate. At some points plates meet at a so-called strike-slip margin, sliding past each other jerkily along transform faults. The San Francisco earthquake was produced by just such a jolt. The only certainty of geologic structure is constant change. "For the first time," Brent Dalrymple has said, "there was really a theory of how the earth works."

Ironically, a treaty during the Cold War has helped back up Wegener's basic theories. Seismic monitoring systems were installed around the globe as part of an agreement among the major world powers to ban nuclear testing. Seismologists are geophysicists who analyze how low-frequency energy waves travel through rock. As vibrations from explosions or earthquakes are reflected and refracted from rocks with different densities or at different levels of the earth, these scientists have learned much

In this map showing the earth's tectonic plates with arrows showing the direction of their movement, three types of plate margins are indicated: The thick lines indicate where plates collide, the medium-width lines show where plates are spreading, and the thinnest lines show the margins where the plates continually glide past each other. The orange dots representing volcanoes correlate markedly with plate margins.

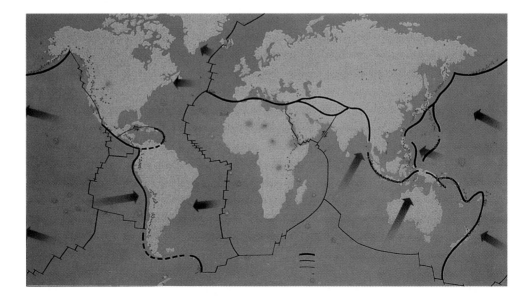

Almost demonized in the United States, the San Andreas Fault in California is a giant shear zone, the area where the Pacific Plate slides northwest along the west coast. In movies and pop novels, this action will eventually sink Los Angeles like Atlantis into the sea, making Tucson a coastal city. In fact, the process will inexorably push Hollywood toward the San Francisco Bay Area, but at an annual speed of only about six centimeters.

about the interior structure of the mantle and core. Those on watch for nuclear tests discovered an unexpected structure stretching along the arc of earthquake zones, a hundred-kilometer slab of rigid material (lithosphere) extending downward into the asthenosphere, or the hot, weak, partially molten part of the mantle.

For the first time scientists realized how exactly the subduction of plates complements the seafloor spreading at the ocean ridges, while convection currents boil up from deep within the earth's mantle. The mantle rises roughly 1,740 miles up from the cove's edge to the lithosphere, its temperatures diminishing from about 6,690 degrees to around 2,000 degrees where it meets the lithosphere. In this mammoth furnace, do great thermal currents power the entire process that drives the spreading of the seas and the life and death of crust at the subduction ridges? Perhaps, but not necessarily. There are still some blanks in the story.

To our ancestors, to our immediate forebears, this picture of the hard earth liquid beneath our feet would seem irrational, impossible. But it is rational analysis of unexpected, puzzling data that produces tectonic theory, seemingly the best possible integration of what science and technology have revealed in the twentieth century about the basic structures of the earth.

Darwin's Long Shadow

At the beginning of our century the majority of scientists believed that the revolutionary ideas of Charles Darwin about the gradual evolution of plant and animal species over the millennia were probably on target. But many religious leaders, along with the general public, were skeptical at best, outraged and inclined to be vindictive at worst. Deeply held traditional beliefs were, for many, an impassable obstacle to dealing rationally with emerging theories about the origins of humankind. Even as the twenty-first century nears, the United States is the only developed country where a great many people who consider themselves educated dismiss Darwinian thought.

It's worth remembering that the basic concept of evolution was in the air before Darwin, but he was the first scientist to begin to explain how the process actually worked. In 1858 he proposed the concept of "natural selection" in his book *On the Origin of Species*. Although added to over the years by theories and discoveries in many scientific disciplines, his basic idea remains scientifically credible.

In short, a species survives because individuals in that species vary in their ability to survive and/or in their ability to reproduce. Darwin saw that these variations, which can be inherited, are fundamental to life: The process of natural selection encourages the survival of individuals who randomly acquire traits that are most appropriate to their environment. A specific trait will vary from individual to individual; if favorable to survival under changing conditions, it will become increasingly common in the species over time. A marking that camouflages an insect from a predator, for example, is favorable to the marked individual. Therefore, similar individuals will become more common than unmarked insects.

In addition, for a number of possible reasons, some individuals are more likely to reproduce offspring than others. Their descendants will become more common in the species. As Darwin explained in *On the Origin of Species*:

> *. . . any being, if it vary however slightly in any manner profitable to itself under the complex and sometimes varying conditions of life, will have a better chance of surviving, and thus be naturally selected. From the strong principle of inheritance, any selected variety will tend to propagate its new and modified form.*

Paradoxically, Darwinian thought came to dominate scientific thought even as it was hotly challenged in the public arena. In 1925 in Dayton, Tennessee, an obscure country town not far from Chattanooga, a twenty-four-year-old high school science and math teacher named John Scopes was tried for teaching evolutionary theory at Rhea Central High School. The judicial circus that followed became infamous worldwide as the Monkey Trial.

Many in Dayton and elsewhere believed that Darwinism was a planned assault upon the Old Testament account of Creation, not a scientific discovery free of any religious agenda. They did not choose to believe, to use the popular misunderstanding of Darwin, that they were "descended from apes."

But the true story behind the trial reveals unexpected motives. Scopes taught evolutionary theory from a biology text that had been approved by a state committee before the legislature passed the Butler Act, which said: "[It is] unlawful for any teacher in any of the universities, normals and all other public schools of the state, to teach any theory that denies the story of the divine creation of man as taught in the Bible, and to teach instead that man has descended from a lower order of animals."

This was a red flag to the American Civil Liberties Union, which publicly offered to defend anyone charged under the provisions of the Butler Act. Seeing a grand opportunity for free publicity for their little town, a group of local boosters convinced Scopes to become the defendant in a test case. The boosters got more than they bargained for.

Clarence Darrow, the great rationalist lawyer and espouser of unpopular causes, arrived on the scene to represent Scopes for free. He opened fire with a typically acerbic sound bite: "[This is] the first case of this sort since we stopped trying people in America for witchcraft."

William Jennings Bryan, a candidate for President three times, was hired as assistant prosecutor by the World Fundamentalist Association. He countered Darrow's opening sally by calling the upcoming trial "a duel to the death between Bible Christianity and infidelity."

The judge knew that the case was neither. It was merely a question of whether or

not a state can constitutionally determine what subject matter is taught in public schools.

But Darrow slyly shifted the grounds of the argument by attacking the patently weak points in the "creationist" view. Could the witnesses for the prosecution agree on the age of the earth, using scriptural references alone? They could not. Could they give a ballpark date for the Great Flood described early in Genesis? No.

In a legendary confrontation with Bryan, Darrow skillfully pitted logic against conviction, using the witness's calculations:

> Darrow
>
> *Do you believe that all of the species on earth have come into being in the 4,200 years since the Great Flood?*
>
> Bryan
>
> *Yes.*
>
> Darrow
>
> *Do you not realize that some civilizations have existed for more than 5,000 years?*
>
> Bryan
>
> *I have never felt a great deal of interest in the effort that has been made to discredit the Bible.*
>
> Darrow
>
> *Do you think the earth was made in six days?*
>
> Bryan
>
> *Not in six days of 24 hours. . . .*

Clarence Darrow, arguing for the defense, turned a case about the constitutional right of the state of Tennessee to mandate a school curriculum into a landmark debate about the scientific accuracy of evolutionary theory. Although it has been modified and refined by discoveries in anthropology, geology, and chemistry throughout the century, this view of the development of life is basic to today's academic science.

Darrow won such skirmishes easily enough, but the outcome of the war was a foregone conclusion. The trial settled neither the legitimate constitutional issue in question nor the viability of evolutionary theory. Religion was neither vindicated nor confounded. The jury of Scopes's Tennessee peers voted for conviction and were widely portrayed as yokels. Scopes was fined a hundred dollars and was widely portrayed as a hero. The conviction was later overturned on a technicality by a state appeals court.

Dayton fell back into obscurity. The Butler Law remained an embarrassment and teaching obstacle in Tennessee until 1967, when it was found unconstitutional.

The Missing Link

Scientists were amused or vexed by this kind of thing, but they pressed on. While Darwinism made logical sense, it would be nice to find the so-called missing link, a skeleton of the creature then imagined to bridge a gap between humans and a supposed apelike ancestor. This concept is now known to be flawed, though the term still lives in some popular writing. Today's scientists speak instead of the "last common ancestor" (LCA). This creature, in theory, would be the still-unknown progenitor of species that are separate in our knowledge and experience, such as humans and chimpanzees. Unlike "missing link," the notion of an LCA does not suggest that human evolution has occurred in a linked, unbroken chain down to the present. Several species of humanlike, or hominid, creatures may have lived alongside each other, but we alone have survived. Differences between other apes and the human ape are several and may have occurred at different times, making nonsense of the idea of a single "link."

But in the nineteenth century, the idea still seemed scientifically sound. Paleoanthropologists, or scientists who specialize in the study of early humans, had been uncovering very ancient bones of hominids since 1856, when the first remains to be recognized were dug up in Germany's Neander Valley. Their brains would have weighed about three pounds apiece, just like ours. So many other "Neanderthals" were found throughout Europe west of the Ural Mountains that it seemed clear by 1925 that the wellsprings of human evolution had to be European. This theory fitted pleasantly with the assumption of many scientists of the time that "Western civilization" was superior to that of the rest of the world. In addition, another fossil, Cro-Magnon man, had been dug up in France. We now know that this creature was a completely modern human being, but the discovery was originally seen as confirmation of Eurocentric evolution: Presumably, the first hominids would have taken the first steps in creating civilization, giving supposedly preeminent European culture the most ancient of human origins.

Half a century before, during the excited debates over Darwinian thought, the influential British scientist Thomas Huxley had argued that humans, or *Homo sapiens,* were obviously a family within the order of primates. He pointed out that the hands and feet of gorillas and humans are more similar than the hands and feet of gorillas and orangutans. It would not be easy to find an unmistakable one-to-one "link"; the relationship would be more like cousinship.

But other scientists hoped to find a single common ancestor. In Trinil, Java, the bones of an extinct ancient upright ape were found. The creature, dubbed Java Man, was accepted as the Asian forebear of all the Neanderthals, who in turn were the progenitors of *Homo sapiens.*

But as Jacques Hublin of the Museum of Man in Paris has explained recently, the ancestral portrait of the first humans was still being conceived in our modern image: more intelligent than our primate cousins from the very beginning; capable of making tools; physically attractive by our standards. They were expected to spring up in dominant Europe or, at worst, in charmingly exotic Asia. Hublin refers to this mind-set as the "Adam complex."

The Skull of a Child

These assumptions were forever shattered when a primate's fossilized skull was uncovered in a Buxton Limeworks quarry in Taung, South Africa. On November 28, 1924, Professor Raymond Dart of South Africa's Witwatersrand University received the skull from the Taung quarry. After a blast during mining operations, it had tumbled out in the rubble, heavily calcified and partially embedded in the cave rock, known as *breccia.*

Dart, an Australian anatomist who specialized in studying the embryonic development of the brain, cared nothing for fossils, but a colleague thought he might be interested. It was incomplete, but Dart recognized some startling features. Although it obviously belonged to a young apelike creature with a smallish brain and protruding jaw, it had a canine tooth, or eyetooth, that was small, rather like a human's, instead of fanglike as in the chimpanzee or gorilla. In other words, it might have used tools rather than teeth to fight with members of its own species. The spinal opening at the base of the brain, the foramen magnum, indicated that the animal balanced on a more or less upright spinal column—that is, it walked on its hind legs. There was still more. Because of where it was found in the rock strata, the so-called Taung skull was very, very old.

The famous Taung skull, discovered in southern Africa in 1924, is probably the remains of a very young hominid killed and devoured by a predatory bird. It was twenty-three years before the discovery of an adult skelton confirmed that these creatures, known as Australopithecus africanus, could walk upright.

The young African, now known to be about four years old when he died, had toddled across the land some two million years earlier.

Today, scientists believe that he was probably killed and eaten by an eagle, for the site in the quarry has the characteristics of a large raptor's nest. The parent brought back portions for its hungry young; then the bones fell into the crevice below. Similarly, most of the hominid skeletal parts found in the caves of South Africa seem to be left-overs from the at-home meals of leopards or other carnivores.

Dart was certain that the "missing link" lay right in his hands. Not only was it the first primate fossil ever found south of the equator on the African continent, but it also seemed to show a metamorphosis between ape and human being. The oldest human ancestor ever located at the time of its discovery, it is now known as *Australopithecus africanus,* from the Latin *australis* for "south" and the Greek *pithékos* for "ape." At age thirty-two, after scraping away with his wife's stainless steel knitting needles at a calci-fied skull for forty days, Dart had made a stunning leap of analysis.

But rejection of this notion was predictable. The Adam complex kicked in. A young striding ape with no proven toolmaking skill did not belong in evolutionary Eden. Dart's findings were published on January 6, 1925, in London, where it was conve-nient for the best-known scientists of the day to read and ridicule them.

One objection was Piltdown, a skull found in England in 1912. Not only had it lived appropriately in Europe, it also had a large brain, thus fulfilling preconceived no-tions in two ways. Most scientists believed that the growth of the brain was a key first step in human evolution. From its discovery, however, Piltdown looked questionable to some clear-thinking observers. For one thing, no similar fossils were ever found. In 1953, what looked so much like the joining of a modern human skull with an ape jaw was proved to be just that: in other words, a surprisingly successful hoax. Nonetheless, some museums still displayed reconstructions of Piltdown a decade later.

Dart fortunately was able to ignore academic scorn and believe his own eyes, logic, and intuition. Not knowing that it would take a quarter of a century of dedicated work and analysis to prove his theories, he and his colleagues set upon the quarry and environs with a passion. In 1936, an adult skull with the child's characteristic features was uncovered forty miles north of Pretoria at Sterkfontein by Robert Broom, a seventy-year-old retired physician. By 1947, deep in the caves there, many more remains of *Australopithecus africanus* had been found, including enough of a skeleton to affirm

Dart's hunch that these hominids walked upright. Then a third site began to yield skeletons. Obviously communities of these creatures had been established in the region long before history. Broom and his colleagues found about one hundred hominid specimens; another five hundred have since been uncovered.

One adjustment had to be made to Dart's original analysis: These hominids were not the missing link being sought elsewhere but part of the human family.

And Older Still . . .

Major discoveries by the famous Leakey family in the 1950s and 1960s achieved clarifications of human lineage and cousinage while continually astonishing the world. Louis S. B. Leakey, born to British missionary parents in Kenya in 1903, began digging in Tanzania's Olduvai Gorge when he was twenty-eight years old. He was at first interested primarily in ancient stone tools, but eventually he and his wife, Mary, virtually stumbled upon unexpected fossil remains of hominids. Their most important discovery occurred in 1959, when Mary discovered *Australopithecus boisei.* Because of its large teeth and jaws, this creature is classified as a robust australopithecine. Its discovery confirmed Dart: In Africa lay the bones of our earliest known ancestors. The Leakeys' son Richard and his wife, Maeve, continued the family paleoanthropological tradition through the end of the century.

In 1974 Yves Coppens and Donald Johanson discovered a partial skeleton of an australopithecine far to the north in Hadar, Ethiopia. The find shows how an apparently modest bit of datum can blow away preconceptions, for it included a pelvis that proved the creature had to walk upright—that is, be bipedal—and could be dated to between three and three and a half million years ago. Assigned to the species *Australopithecus afarensis,* this female skeleton became world-famous as "Lucy," reportedly because the band of scientists celebrated their spectacular find with a party where the recorded music included "Lucy in the Sky with Diamonds," a hit by the Beatles. With about 40 percent of her bones found, three-foot-tall "Lucy" is one of the most nearly complete fossil hominids yet uncovered.

In the Olduvai Gorge in Tanzania, essential elements in the story of humankind's beginnings have been found in the last half century. Accidents of preservation have probably distorted the record, but individual finds like the skull of Australopithecus boisei *(inset) suggest a variety of ancient hominid forms that was unimaginable even five decades ago. Found in 1959 by Mary Leakey and sometimes identified as* Australopithecus robustus, *the creature gained instant worldwide fame as probably having lived almost two million years ago.*

Four years later Mary Leakey discovered footprints of two walking creatures that had been made about 3,600,000 years before in the South African desert at Laetoli, about twenty-five miles southwest of the Olduvai Gorge. This was an astonishing site with trackways made by hundreds of other individual mammals and birds—in sum, the documentation of several weeks of activity in the area. The prints were indelibly preserved because of a rare combination of factors: Volcanic ash with a particular chemical composition fell at just the right rate in conjunction with just the right amount of rain to produce a substance that became cement-hard at the time and endured for aeons.

In 1984 Alan Walker and Kamoya Kimeu found the 1,800,000-year-old skeleton of *Homo erectus* on the western shores of Lake Turkana in Kenya. In this case the lead of the story was not great age but closer kinship. Like the skulls of Neanderthal and Java Man, this one was oval and had very prominent brow ridges. Walker's *Homo erectus* was almost a complete skeleton, an unusually rich find. Death had been caused by a tooth abscess. Apparently, the dying hominid had made its way to a swampy area, much as other wounded animals will seek out a water resource as they weaken.

First visited in modern times by Richard Leakey, the area around Lake Turkana is dismayingly inhospitable. The lake water is too saline for most people to drink comfortably, highly venomous snakes slither into campsites, and some nearby tribal groups are hostile. Scientists have literally given their lives to pure research in this region. Still, archae-

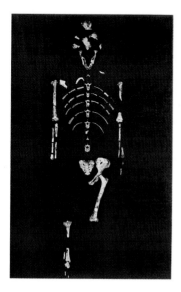

Far left: *About 3,600,000 years ago, two upright adult hominids, most likely* Australopithecus afarensis, *walked down this seventy-meter stretch of pathway in Tanzania, leaving the oldest known footprints of bipedal walking. Expert readings of this remarkable ancient saunter show that the adults had well-developed arches. To the right, an extinct three-toed horse trotted at some point in a scene recorded by a rare combination of weather and the drifting of volcanic ash. In 1997, the oldest known footprints of an anatomically modern human, fossilized in sandstone 117,000 years ago, were found in South Africa. Left: The celebrated "Lucy" is important to science not only because of her age, about 3,500,000 years, but because her remains are phenomenally complete, about 40 percent of the entire skeleton.*

ologists are lured by the variety of materials already found in the exposed sediments, some of them two million years old. Stone tools found by Glynn Isaac and his colleagues added to the hominid archaeological record by acting, in Isaac's term, as "calling cards." This indication of early hominid travels has sparked new studies into their territorial behavior.

Why do so many gaps remain in the fossil record? Briefly put, because geological coincidence has favored the preservation of some populations over others. Probably various types of hominid were scattered over the entire continent of Africa, isolated geographically from each other and adapting to distinctive local habitats. They evolved and eventually died out, and their bones became dust. But in areas like the Eastern Rift Valley that runs south from Ethiopia through Kenya or in the caves of South Africa, sediment accumulated in layers above hominid remains, thus creating fossil-bearing deposits. The bones of the hominids who died in those areas provide a dramatic but necessarily partial view of the complex development of the human family.

Thousands of Peas

If these trends in paleoanthropology tended to vindicate Darwin's ideas throughout the twentieth century, other discoveries provided ways of thinking about evolution that he had not imagined.

How did creatures adapt to change? How did traits that promoted survival get passed down to the next generation? At almost exactly the same time Darwin was setting down his ideas in *On the Origin of Species,* an Augustinian monk was coming to grips with the mathematical basis of genetics in a monastery garden in Brno, Moravia. Gregor Johann Mendel, a farmer's son who joined the religious order in 1843, had been encouraged by his abbot to pursue his keen interest in the natural sciences.

After studies at the University of Vienna, Mendel began rigorously experimenting with plant breeding in 1856. He worked with pure strains of various kinds of garden pea, crossbreeding several thousand plants to produce hybrid progeny. He developed a new technique for controlling pollination of his plants, thus ensuring that he knew exactly which plants cross-pollinated. Taking note of seven specific characteristics, he kept meticulous records of the rate of their appearance in subsequent generations.

For example, if one parent plant had wrinkled seeds and the other smooth and rounded seeds, the next generation would produce nothing but smooth seeds, while in the following generation about a fourth of the seeds would be wrinkled. Eventually Mendel realized through his statistical analyses that this pattern of inheritance revealed that one characteristic appeared to be "dominant" over the other, which in turn appeared to be "recessive." The same pattern could be discerned in each of the seven distinguishing features he studied; the "dominant" of a pair of characteristics appeared in the first generation, and the "recessive" trait reappeared in about a quarter of the second generation. Moreover, the rate of incidence of one characteristic—say, the texture of the skin of a pea—moved through succeeding generations independently of the appearance of another characteristic, like color.

What were the elements that carried these features? Mendel didn't know exactly, though he called them "factors," but his tables of inheritance patterns were, and are, models of research documentation. He recognized that two of the hereditary units were contributed by each parental gamete, or reproductive cell. In other words, parents separately contributed hereditary information to their offspring. Previously scientists had assumed that such information blended as a kind of hereditary soup. The monk's work showed that it was transmitted as discrete pieces that interacted with one another in predictable patterns.

By 1865 Mendel felt ready to publish his findings, but his work was unappreciated by academics from Vienna to Cambridge, Massachusetts. In his own country, which was then part of the Austro-Hungarian Empire, there was tremendous interest in improving sheep breeds in order to boost the wool trade. Evidently no one grasped how successfully Mendel's work with the humble pea could be translated to that potentially more profitable sector of the economy.

In 1868 Mendel was chosen abbot of the monastery. From then on he concentrated on the work of the spirit. His great contribution to the odyssey of science was not rediscovered and understood until the twentieth century. Ironically, his insights into the properties of inheritance had been understood by people of the soil for millennia. As Darwin had pointed out, farmers selectively breed their livestock and plants in a process he called "artificial selection." In the Old Testament, wily Jacob grew wealthy by choosing the strongest animals from his father-in-law's sheep and goats and breeding them. Certain kinds of science spring almost whole, it seems, from the humblest experiences of daily life, but not until someone subjects "common sense" to rigorous analysis.

Morgan's Fruit Flies

By 1900 it was now known that the chromosome, a rodlike body found in the nucleus of a cell, had something to do with inherited characteristics. Chromosomes, which are composed of protein and nucleic acid, occur in pairs in animals. Two sets of these chromosomes, one set from each parent, are found in a fertilized egg. In other words, a parent can pass along only half of his or her genetic material.

Working at New York's Columbia University, Thomas Hunt Morgan had first been wary of both Mendel's and Darwin's work, but *Drosophila melanogaster*, a fast-breeding fruit fly, caused a dramatic shift in his thinking. Short-lived and enthusiastically reproductive, drosophilas have only four pairs of chromosomes, making it comparatively easy to track their expanding generations. One female will easily produce two to three hundred offspring. Since the life cycle is a bit less than two weeks, the lab could see at least twenty-four new generations of flies each year. Moreover, they could be grown cheaply on mashed bananas.

Morgan's lab at Columbia, thanks to his energetic subjects and an unusually large number of hardworking graduate students, took on the tempo of an industrial project. Not only did the flies breed in legions, but they also tended to throw up a significant number of mutations—i.e., new traits in the evolving population of *Drosophila*. Such characteristics are not inherited from either parent fly. They are radical changes that die out if they are inappropriate to the environment, or they become established in the species if they are advantageous. In other words, along with Mendel's tables, Darwin's theory of natural selection was being affirmed by the fertile fruit flies.

In January 1910, for example, a white-eyed male fly appeared among his fellows, which typically had red eyes. Morgan collared this mutant and bred him. As Mendel had predicted, the next generation all had red eyes, just like their normal-appearing mother. In the third generation, again following Mendel's tables, one in four flies had the unusual white eye.

Eventually researchers saw that each chromosome contained hereditary material in the form of genes. The number and arrangement of these genes are specific to a species; no other creature on earth carries the same genetic information as *Drosophila*. The genes determine the traits of an organism.

But why do mutations occur? Morgan's colleague Hermann Muller found that bombarding the flies with X rays seemed to produce mutations in succeeding generations. He saw that all mutations, whether induced by outside factors or occurring spontaneously, resulted from chemical reactions in the genes.

Scientists correctly assumed that the lessons of the flies applied to all other organisms.

We Are Chemicals

Halfway through the twentieth century science knew an enormous amount about chromosomes, but not enough. The pieces did not yet fit together. A molecule called deoxyribonucleic acid (DNA) had been found in the nucleus of the cell; evidently it played an important role in inheritance. Uncoiled and pulled straight, the amount in a single cell would be about eight feet in length. In 1951 Dr. Linus Pauling showed that some pro-

In close-up, the most famous fly in science, Drosophila melanogaster, *exhibits a certain presence here. Beginning in 1907, Thomas Hunt Morgan and his colleagues encouraged this fruit fly to follow its inclinations, i.e., to breed at a rapid pace, again and yet again. Such enthusiastic procreation made possible Morgan's understanding of the basic mechanisms of heredity.*

teins, a key substance in the production of all living cells, were shaped like a helix, or coiled strand. It was also known that genes, whatever they were, somehow controlled the manufacture of proteins in each of the human body's three trillion cells.

In 1952 researcher Rosalind Franklin made the first X ray of a molecule of DNA. It suggested that like protein, DNA might have a helical structure, but this theoretical model did not fully explain how all the known elements of DNA could fit together, much less reproduce themselves.

Francis Crick and James Watson were intrigued with this conceptual problem. In a flash of insight Watson realized that Pauling had erred in his calculations for the shape of protein. The only structure that could explain DNA was a double helix, two strands winding around each other. The strands are rows of nucleotides, and the nucleotides of one strand are bonded with hydrogen to nucleotides on the second strand. Each nucleotide contains one of four bases: adenine, guanine, thymine, and cytosine.

At this point the nature of science in the twentieth century was about to undergo a radical change. From now on genetics would not be the province solely of biology and

Right: *Critical to the discovery of the structure of genetic DNA was the famous "exposure 51," an X-ray diffraction obtained by Rosalind Franklin in May 1952 from a fiber of DNA in the thymus gland of a calf. "The instant I saw the picture my mouth fell open and my pulse began to race," James Watson said later. He realized that her photo implied that the structure, seen from above, had to be an intertwined double helix.* Far right: *Postgraduate students James Watson* (left) *and Francis Crick built upon ideas of protein structure developed by Linus Pauling to solve the problem of the structure of DNA.*

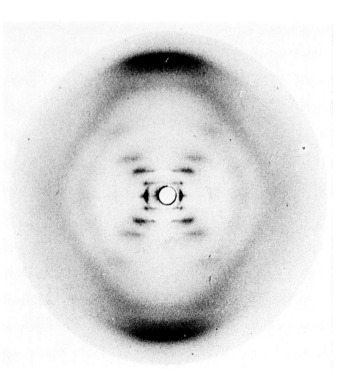

biologists; physics and chemistry would join a new multidisciplinary approach to the problem of the nature of organic life.

When Watson and Crick built a model of DNA, they could see concretely how information about inherited traits is passed on. The two strands mirror each other: Thymine and adenine bond only with each other; cytosine and guanine are also complementary. Therefore the strands can copy themselves, or replicate, by pairing bases. Opposites attract continuously, uncoiling and pairing in the dance of life.

It is nearly impossible to overestimate the significance of this discovery, for it altered the structure of scientific thought even as it illuminated the structures of all life. As Michael Lerner has said, "In science and other intellectual endeavors, there are two kinds of people: the architects and the bricklayers. There are a lot of bricklayers, and they do essential work, but Watson and Crick are among the few real architects of science."

The mechanism of DNA replication is universal throughout the living world: *Drosophila*, garden pea, *Homo sapiens*. Darwinian evolution turned out to be a matter of

chemistry. Molecules are in charge. It turns out that human beings and our closest relatives, chimpanzees, share 99.6 percent of our active genes. The distinguishing 0.4 percent is not evidence that we somehow soar above the rest of creation, as many people believed at the outset of the century.

The First Sparks of Life

As the century progressed, many of the startling discoveries of the early decades were affirmed and reaffirmed. New information only extended all boundaries. The oldest rocks on earth, found on the west coast of Greenland, are 3.8 billion years old. But that is not the whole story. Using the ticking rocks of radioactive dating, Caltech's Clair Patterson in the 1950s turned her attention to meteorites. It turned out that the lead in this extraterrestrial debris, when compared with the lead in volcanic lava and the sediment of the seas, reveals that the earth and indeed the entire solar system have to be from 4.5 to 4.7 billion years old.

By comparison, Darwin's work implied that the evolution of all life could be taken back with some certainty for about 550,000,000 years, long before any close human cousins began ambling across the earth. But we now know, life has been present since the geologic era known as Precambrian time, which began at least 3,800,000 years ago. By then the permanent crust of the earth had formed, and the great engines of sedimentation and erosion were beginning to produce their continuous alteration.

Conventional optical microscopes cannot detect fossils of the earliest types of life, which are extremely small. Researchers turned to the electron microscope, a device developed in the 1930s and widely available for laboratory use by 1939. A beam of accelerated electrons has wavelengths that are perhaps a hundred thousand times shorter than light waves. The beam illuminates an infinitesimal subject; then the scanned image is hugely magnified with an electromagnetic or electrostatic field and reproduced on either Polaroid or standard photographic film. Today's electron microscopy can capture images a million times the size of an object or scan in three dimensions. The largest device yet built is three stories tall and uses three million volts.

Back in the mid-1960s electron microscopers searching through microscopically thin sections of Precambrian rock realized that certain rock masses called stromatolites are in fact structures built by colonies of bacteria that lived as long as 3,600,000,000 years ago. Evidently the top layer of each mass holds the remnants of cyanobacteria, a type of bacterium that can survive in dim light and live by consuming oxygen.

Astonishingly, stromatolites still thrive in some parts of the globe as our contemporaries, for cyanobacteria and their relatives are able to live out their rudimentary lives in a variety of diverse conditions: from fields of ice to acid hot springs, from the seawaters rich with chemical salts to crevices in barren rocks. It is they, before other forms of life could evolve, that "learned" to photosynthesize—that is, they manufactured carbohydrates from the water and carbon dioxide of Precambrian time, expelling free oxygen (O_2) as a by-product.

The process was slow, for the earth is vast. The increasing availability of oxygen in the water enabled the cyanobacteria to evolve into larger, multicelled forms of blue-green algae, perhaps about 1,850,000 years ago. The diversity of life, in geologic terms, had begun to accelerate. Only 200,000,000 years later the algae exploded into many different forms, becoming the ancestors of jellyfish and seaweeds.

Evolutionary biologists have come up with crisp images that memorably demonstrate the oceanic reaches of the earth's history. For example, if imagined in terms of miles, the age of the earth could be considered equivalent to ninety-five hundred miles, while humans popped up nine miles back. If the earth's current age is represented on the scale of a calendar year, the cyanobacteria began their colonies some time in April, and our earliest ancestors slip in at about five minutes before midnight on December 31.

Yet even these ancient bacteria may not have been life in its first form. In 1979 a tiny spherical submersible named *Alvin* allowed scientists to photograph undersea black smokers for the first time. Found deep within the oceans near the regions of volcanic activity along mid-ocean ridges, these undersea gouts of superheated water with their dissolved minerals and gases can be as hot as 662 degrees Fahrenheit.

In these black smokers, as in steaming geysers aboveground in Iceland and in oil wells dug deep toward earth's molten center, Dr. Karl Stetter of Woods Hole has found heat-loving bacteria called hypothermophiles. He boldly theorizes that similar submicroscopic organisms thrived aeons before cyanobacteria, somehow eking out an existence on a dark, fiery, poisonous mass of active rock that bore little obvious resemblance to our own temperate, fertile earth.

Prebiotic chemistry, the study of natural chemical reactions that occurred before life began, has only recently become a contentious field. In 1953 graduate student Stanley L. Miller performed an experiment that became the textbook explanation for the origin of life, the so-called soup theory. In a glass vessel he combined water with methane, ammonia, and hydrogen gases, thus attempting to mirror the earth's ocean and the sup-

posed composition of its early atmosphere. Then, mimicking a bolt of lightning striking the primordial soup, he sent sparks of electricity into the mix. Days later he discovered amino acids, the building blocks of life that are necessary to form proteins, as well as other organic chemicals within the vessel.

This was a classically elegant demonstration, but it left an important question unanswered: How did the chemicals combine to form the first molecules of living organisms? The answer has not yet been found. Furthermore, the accepted picture of the earth's early atmosphere has changed: It was probably O_2-rich with some nitrogen, a less reactive mixture than Miller's, or it might have been composed largely of carbon dioxide, which would greatly deter the development of organic compounds.

Some theorists are attracted to the possibility that the building blocks of earth life did not originate here but arrived along with the asteroids, meteorites, and comets that continually bombarded the planet during its first billion or so years of existence. Others believe that tiny dust particles traveling through space—and still falling to earth at a mass some one hundred thousand times greater than the mass of meteorites that reach the ground—might have brought along the organic compounds that generated the first life here.

In the late 1990s other scientists began to argue that life may have originated near deep-sea vents like black smokers, where carbon monoxide, hydrogen sulfide, nickel, and iron pyrites all would be found. According to Dr. Gunter Wachtershäuser, a Munich chemist, reactions that are essential to the beginning of life can occur on ferrous sulfide. For example, the gases found at black smokers will form acetic acid when in the presence of a mixture of ferrous and nickel sulfides. The carbon atoms are chained together naturally, much as living cells structure themselves.

To have several competing theories of the origin of life is no new thing, but the areas of agreement are unprecedented: that life began as a simple chemical reaction, that the process started more than 3,500,000,000 years ago, that the nature of the prebiotic atmosphere of the earth is essential to understanding the beginnings of the DNA that makes us what we are.

Thought to be extinct until only four decades ago, the stromatolites seen here in Shark Bay in western Australia form "turfs" near the beach. A fossil record of cyanobacteria, the bacteria that probably generated the oxygen in the earth's atmosphere aeons ago, some stromatolites are apparently 3,600,000,000 years old, or only three tenths of a billion years younger than the oldest known rocks. They have so far been found only in Australia and Jamaica.

Refining the Data

In this century we have learned that the origins of life and of our earth are almost unimaginably ancient, and we are beginning to learn that these two beginnings have perhaps always been intertwined.

What more can we hope to learn?

There are still mysteries to be solved about the traveling of the continents. Is it just coincidence that all the visible surfaces of the earth, if joined together, would make a complete globe about half the diameter of the earth we know? A few scientists speculate that it has mysteriously expanded over time; in their view, the absence of a subduction zone near the coast of Antarctica indicates that the frozen continent and neighboring Africa remain stationary while the rest of the earth's crust slides off to the north, continually expanding. Most traditional scientists discount this view—indeed, there are polar glacial markings in West Africa—but the history of discounted views in science is greatly mixed. Perhaps answers to this mystery and others will be solved by satellite gravimetry, which will place gravimeters into orbit in order to map the seabeds with unprecedented precision.

There is no generally accepted explanation of the great leap taken by *Homo sapiens,* making us the only apes that think in symbols and communicate in language. The human brain is four to five times larger than the brains of our nearest ape relatives today, but more significant are the structural differences: The prefrontal cortex has grown, and various areas of the brain can be taught to think symbolically. What happened? Why?

The search for fossils continues in South Africa, East Africa, and even into Chad in West Africa at something like fever pitch. Using such new methods of dating as electron spin resonance, which can count the electrons trapped in tooth enamel with microwave radiation and thus estimate how long it has been exposed to natural radiation damage, anthropologists are arriving at increasingly accurate ages of old and new finds. The science of molecular evolution, which includes studying the DNA of humans and their ape cousins, can extrapolate how we all may have been chemically related and

At White Island in New Zealand, sulfur fumes boil out of the vents of an active volcano. This lethally beautiful scene recalls the birth pangs of the earth, when sulfurous volcanic activity made only the simplest, most tenacious forms of life possible. The story of atmospheric change that made life possible, though still incomplete, is being solved and, of course, debated by physicists and chemists.

when. Currently the molecular experts infer that hominids and chimpanzees had a common ancestor—not a "missing link" in the discredited sense—between 5,000,000 and 7,000,000 years ago. Maeve Leakey, Richard's wife, has dug up an *Australopithecus anamensis* as much as 4,200,000 years old in northern Kenya. Tim White, working in southern Ethiopia, has found a 4,400,000-year-old hominid.

Meanwhile, the entire complex structure of DNA, it now seems, will be charted by century's end, if not sooner. In a three-billion-dollar concerted scientific probe known as the Human Genome Project, not only will the more than one hundred thousand genes in human cells be identified, but each will also be placed in its proper sequence among the individual chemicals that make up the map of the human genome,

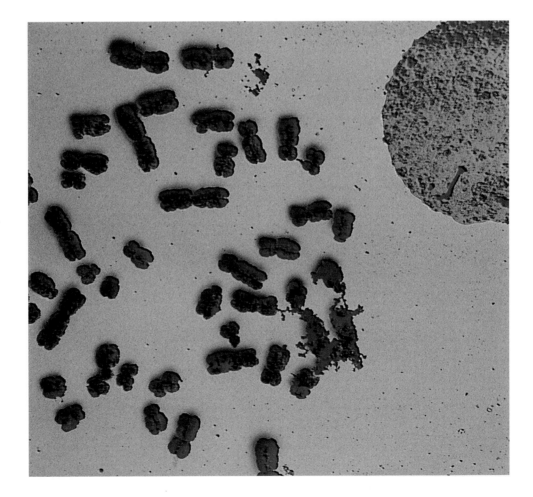

three billion chemical bases. (Put another way, that's a dollar per base in research costs.) This will be a string of letters—*T* for thymine, *C* for cytosine, *A* for adenine, *G* for guanine—that can be imagined as filling 510 volumes of a family encyclopedia. The word "genome" is used as an umbrella term for all the genes in any individual organism.

The goal is to uncover, in James Watson's words, "the complete set of instructions for making a human being." He was the first head of the project when it began on January 3, 1989, under the dual sponsorship of the National Institutes of Health and the U.S. Department of Energy. Scientists in Europe and Asia joined the effort. To avoid destructive competition, the work was divided among researchers throughout the genetics community. The first goal was to map the so-called index markers, which are typical sequences to be found every ten to fifteen million bases, then to look for the more complex sequences in between.

The genes implicated in several hereditary diseases have already been isolated, but this is only the first step. Our genetic heritage includes three thousand separate genetic disorders. Eventually the project may reveal, like a kind of evolutionary clock, exactly when our first human ancestors appeared on the scene by determining when significant changes in DNA occurred. Meanwhile, perhaps in response to eugenics and other misguided genetics policies of the twentieth century, some observers have expressed alarm at the potential misuse of the Human Genome Project. Complete knowledge of an individual's genome, which might include various easily identifiable genetic diseases of body and brain, could become the pretext for discrimination.

But the odyssey of science frequently requires that all of us ensure that breakthroughs in understanding are first of all understood as the servants, not the masters, of human destiny. A genetic map is not in itself evil or beneficent; it is merely, and breathtakingly, the chart of millions of years of evolutionary history and the present and future of one unique organism. In this project and elsewhere the search for origins has revealed unexpected vistas throughout our century. We have no reason to suspect that we have seen more than the first glimmers of the full sweeping epic.

Magnified here 420 times in a scanning electron micrograph are the paired chromosomes of a human white blood cell in metaphase; these chromosomes will split into two groups, forming two exact replicas of the original cell as daughter cells. At upper right is the nucleus of a cell at rest in interphase, the normal condition. During metaphase, a single error in the two trillion chemical reactions required for replication will cause a mutation.

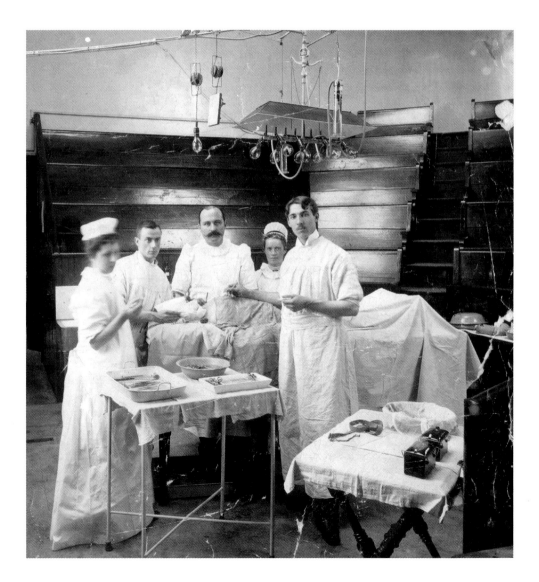

CHAPTER FOUR

MATTERS OF LIFE AND DEATH

On September 6, 1901, in the midst of a public reception at the Pan-American Exposition in Buffalo, New York, William McKinley was shot twice in the abdomen by an anarchist. Because the wounded President bled profusely, the attending surgeon, Dr. Matthew Mann, decided to operate immediately at the exposition hospital rather than risk moving him to the city's better-equipped General Hospital.

Naturally, even in this tense crisis, the surgeons took the few precautions that had become common only within the last two decades: They scrubbed their hands with disinfectant, then rinsed them in a caustic solution of mercuric chloride. They donned surgical masks, caps, and gloves to help prevent the spread of infection.

At 5:39 P.M., as dusk fell, the surgical team began probing for the bullets lodged somewhere in McKinley's abdomen. Shadows deepened, but the gas lamps of the operating room could not be lit; the patient had been put to sleep with ether, a flammable anesthetic. Using a hand mirror, the President's personal physician guided the last rays of the setting sun toward the incision. Under these conditions, it is not so surprising that the bullets could not be found even after twenty-one minutes of frantic searching. Giving up, the surgeons irrigated the wound with a hot salt solution, then sutured it closed with a double row of fine black silk. Off in the distance, too far away to be helpful, the exposition's 409-foot-tall Tower of Light blazed with thirty-five thousand electric lightbulbs, a symbol of technological hope for the future.

Another product from Thomas Edison's laboratories was also on display nearby,

Almost two decades before he would try to save a president's life, Dr. Matthew Mann poses with his operating team beneath the first electric lights ever used in a Buffalo, New York, operating room.

but the newly invented X-ray machine was not used to locate the elusive bullets. Dr. Mann explained later: "To have used the X-rays to simply satisfy our curiosity would not have been warranted." Indeed, it would be a decade or more before X rays would be routinely incorporated into medical practice.

Over the next eight days the President inexorably weakened from infection caused by the bullets hidden within his abdomen. The doctors tried various remedies, including soap enemas to purge the body, then doses of strychnine, whiskey, and camphorated oil. On September 14 McKinley died from gangrene.

But that was to be expected. Survival after a shooting, or many other physical traumas or infections, was generally considered to be a matter of luck and circumstance and prayer. Because of a high rate of infant mortality, perhaps one fourth to one third of all children, the life expectancy for the average American citizen was only forty-seven years, or thirty-five years for an African-American woman. Physicians could do little more than take a patient's temperature, look at the throat, and check the knee reflex. For seriously ill patients, they might provide pain relief with opiates.

Hospitals did not offer effective treatment. Most were charitable institutions set up for the poor, who lacked access to the preferable situation of care in a clean, comfortable home. All too often such hospitals became seedbeds of infection, possibly causing more fatalities than the patients' original medical problems. Isolation was the most trusted method of preventing the spread of a contagious disease, but it was rarely effective.

Disease was a fact of daily life throughout the world. In the United States 140,000 people were killed each year by tuberculosis alone. Rabies was invariably fatal. Typhoid, yellow fever, and influenza waxed and waned in horrifying epidemics throughout the world. In short, no pill, no operation, no preventive technique was widely reliable. Virtually any disease a hundred years ago was a legitimate cause for concern. Epidemics were especially terrifying, even though they were known to be frequent occurrences throughout human history.

Today, we can see that the medical odyssey through the twentieth century has been an accumulation of knowledge on three major fronts: Researchers have learned how various microorganisms that are agents of disease mount their attacks, how chronic illnesses infect or wear down the body, and much about how our internal mechanisms work or break down. Often, advances in this kind of knowledge have made possible the development of bold new surgical interventions, effective antibiotics and other curative or life-prolonging medicines, and reliable preventive vaccines.

These are dramatic achievements, but the most significant improvements in health care in our century began with the understanding that sanitation, clean water, and good hygiene can help prevent devastating epidemics. One such epidemic struck in San Francisco in 1900, a Year of the Rat on the ancient Chinese calendar.

A Plague from the East

When Chinese merchant Chick Gin was found dead of an unidentified illness in the basement of a cheap Chinatown hotel at age forty-two, he indelibly entered U.S. medical and social history. The alert San Francisco medical examiner immediately sent the immigrant's infected glands to the bacteriological lab on Angel Island, the immigration center in San Francisco Bay. The report was swift and ominous: Gin was perhaps the first victim ever to die of bubonic plague on American soil.

This was a scourge associated with the past. In the fourteenth century the so-called Black Death had erased one fourth to one third of the population of Europe in

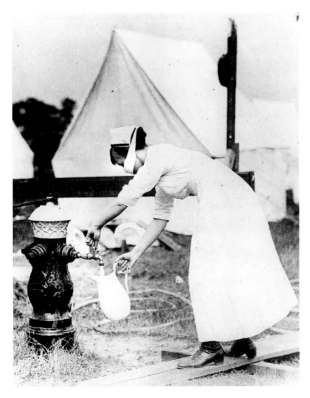

Masked to deter infection from influenza, this U.S. Army nurse is working at a special camp at Corey Hill Hospital in Brookline, Massachusetts, during the epidemic of 1918. Although the devastating disease was not susceptible to cure, physicians had learned how to deal with contagion. First noted in 1414 in France, influenza is still one of the world's most insidious, cleverly mutating diseases.

only two horrific decades. Since the seventeenth century, however, it had virtually disappeared in the West, becoming literary metaphor, not likely threat. But just six years before Gin's death its latest reappearance had killed 100,000 people in China. In India the following year 1,300,000 died. Despite centuries of experience with this devastating illness, perhaps first recognized and studied in ancient Greece, neither prevention nor cure had been discovered. As in ancient times, half or more of infected people died excruciating deaths with chills and high fever, painfully enlarged lymph nodes (especially in the groin area), and perhaps hemorrhages that became appallingly black.

Ignorance was exacerbated by prejudice in San Francisco. Long before Gin's death the established white population of the city had come to disdain Chinatown as a hotbed of yellow fever, leprosy, and poverty. In a report promulgated by the city's board of health, Chinese immigrants had been described in harsh terms: "Alien to our religion, alien to our civilization, neither citizens nor desiring to become so, they are a social, moral and political curse to the community." One day after the diagnosis on Angel Island the fifteen square blocks of the teeming neighborhood were roped off—though only after the white residents there had been hastily evacuated—and twenty-five thousand Chinese were prevented from leaving by guards.

Conventional wisdom had been confirmed—to wit, the Chinese were prone to infection because of their supposedly unclean living conditions. Physicians suspected that their rice-based diet provided too little protein to ward off disease. Finally, the general public opined that the plague arose because of the bad moral character of this immigrant neighborhood, combined with the "miasmic" clouds of filth and dirt that wafted about the impoverished district. Indeed, once confined, the plague now began to spread within its guarded precincts.

The attitude behind this self-fulfilling prophecy had earlier created Angel Island and other immigrant stations. To defend U.S. shores from diseases from abroad, potential immigrants were brought to the island in the center of the bay, stripped naked, and

In 1901, the most vocal residents of San Francisco made clear their conviction that the city's Chinatown had to be responsible for the spread of plague. Had the contagion come from abroad? Probably. Had the congested, generally impoverished living conditions of the Chinese immigrants contributed to the spread of the contagion? Possibly. But neither of these factors was associated with the national, ethnic, or racial background of the residents of the afflicted quarter.

deloused. The slightest mental or physical abnormality, including deafness, a hernia, noninfectious cancers, or even a "sickly pallor," could be deemed sufficient cause for summary rejection.

But the greatest fear of the authorities was infectious disease. A potential immigrant suspected of contagion might be held for observation for weeks, even months. The pain, the deep humiliation of this open-ended isolation, are memorialized today in poetic outpourings of the heart carved on the walls, carved in a veritable prison

on a rocky island surrounded almost entirely by the shores of the land of high promise.

Worse yet, the precautions failed.

Purifying the Source

"Clear the foul spot from San Francisco and give the debris to the flames," an editorialist for one local newspaper growled. A year earlier in Honolulu panicked officials had accidentally done just that, causing a fire that destroyed the plague-ridden Chinese commu-

At Angel Island near San Francisco, quarantine officer Joseph Kinyoun (right) and other officials carefully inspect a potential citizen. The aim was beneficent: to prevent the introduction of devastating epidemics into the United States. From today's point of view, arguably, the results were often cruel. The examination techniques of the day could not always distinguish between the infected and the foreign, the contagious and the terrified.

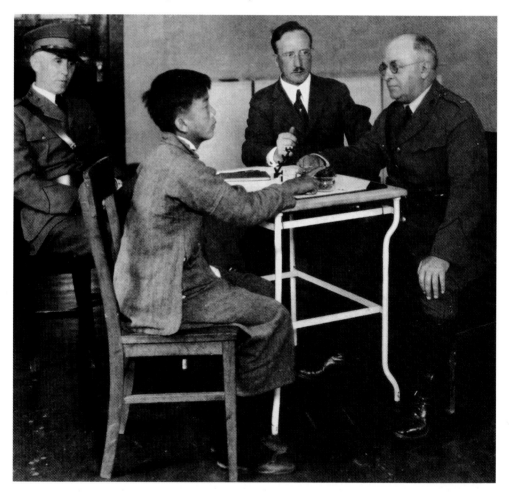

nity there. Hoping to forestall such a violent remedy and also to escape the disease itself, San Francisco's Chinatown residents began hiding ailing relatives and corpses in underground tunnels.

The most measured, informed response to the epidemic originated with Joseph Kinyoun, the quarantine officer who identified the cause of Gin's death. Working briefly in Germany at the world-renowned laboratory of Robert Koch, who had proved that bacteria were one cause of disease, Kinyoun had actually seen the bubonic bacterium under the microscope.

Still, like most other public health experts of the day, he had an imperfect understanding of the plague's tactics and strategies. He believed that the bacteria were emitted by infected human or animal victims, then thrived in the air, soil, or food until they seized upon new prey. By this reasoning, it followed that the disease could be attacked by scouring the affected environment clean.

To purify the streets and structures of Chinatown, the police sprayed everything in sight with sulfur dioxide fumes and mercuric chloride. Buildings that were considered particularly likely to host the contagion were pulled down with crowbars.

To cleanse the people, Kinyoun required anyone who had come into contact with a victim of the plague to submit to a painful serum shot. Everyone else was forcibly inoculated with an experimental vaccine. Neither of these nostrums had been proved effective, nor had possible side effects been studied. Soon many of the inoculated Chinese fell ill.

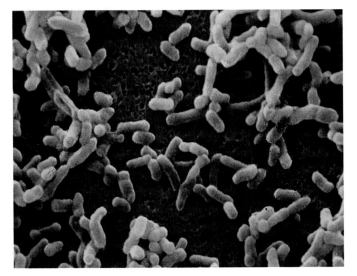

In this false-color scanning electron micrograph, the bacteria that cause bubonic plague in humans appear to be teeming, and so they are, rapidly infecting the host and often causing death. The first known outbreak occurred in Constantinople in A.D. 541 and lasted for three years. At its peak, that plague killed ten thousand victims every day. History's second great plague pandemic began in western Europe in 1346; the third took root in China in 1855.

Thus victimized by more than bacteria, the immigrant community began fighting back during the second month of the quarantine. Since white residents of Chinatown had been allowed to leave and had never been coerced into accepting "treatment," the Chinese brought suit against the city for racial discrimination. They won, primarily because the city was losing business as the result of a flood of unfavorable publicity.

Yet in comparison with notorious plague outbreaks of the past, the San Francisco epidemic was more feared than fearsome. A white victim died on August 11, but the disease still concentrated on the Chinese community, taking a life every ten days to two weeks. By 1904 some two hundred deaths had been reported, although historians believe that many more died and were hidden away.

The Earth Weighs In

In the end this was not a plague that swept through the population causing numerous fatalities, and it gave up no new clues to its origins or cure. Then a completely different kind of natural disaster leveled the city.

The devastating San Francisco earthquake of April 18, 1906, caused more than a thousand deaths and ignited fires that destroyed twenty-eight thousand buildings. Overnight the sparkling jewel of the Pacific became a squalid, filthy refugee camp, with hungry and disoriented rats swarming through the smoking debris. Once again bubonic plague struck.

Fortuitously British scientists working in India had recently discovered the bubonic bacillus in the blood of rats, then in fleas that fed on rats. Suddenly the plague's method of action seemed clear. It did not travel from human to human. Rather, fleas bit infected rats or other animals, then passed on the disease by biting human beings. Kill the rats, and the disease would be stymied.

There was a welcome social dimension to this discovery, as historian Günter B. Risse has noted: "Bubonic plague was now seen as a general menace, independent of race and class. So the primary enemy was non-human: the rat and its fleas, a far less problematic target than stigmatized minorities such as the Chinese."

Led by the U.S. Public Health Service, a kind of War on Rats killed about a million of them. Buildings were ratproofed, garbage was covered, and a new sewer system built. Two months after the beginning of this campaign no new cases of plague were reported. Because of these concerted efforts based upon effective science,

From atop Russian Hill, survivors view the devastated city of San Francisco in the aftermath of the 1906 earthquake. Battalions of rats, confused by the tremors, swarmed through the streets below, carrying bubonic plague. Only eight years before, a physician working to stem a pandemic of bubonic plague in Bombay had recognized that it was carried by fleas on rats.

the death toll for this second outbreak of bubonic plague ended with victim no. 190.

This was a stunning triumph for bacteriology and for the development of public health measures. It showed that, in some cases, a disease could be defeated even though no cure was available. In other words, an alternative approach to good health, funded and administered by government agencies, had been added to the traditional pathways of treatment.

No treatment was available to attack the plague bacterium. Instead, the citizens of San Francisco joined enthusiastically in efforts to rid the city of rats, thereby stifling the epidemic. The recognition that killing rats would prevent an epidemic was a landmark triumph for public health initiatives. Today, the plague survives in some parts of the American West, where it is carried by prairie dogs.

Disease of the Four Ds

Yet another pathway was discovered by Joseph Goldberger, a quarantine physician and epidemiologist with the U.S. Public Health Service back East. Born a Hungarian Jew, he had undergone the degrading experience of being stripped naked and disinfected at New York's Ellis Island as a child. Unquestionably brilliant, Goldberger risked his life several times in research during his first fifteen years of public service, contracting typhus, dengue fever, and yellow fever. Such dedication rarely goes unpunished, and he found himself pressured to head the Pellagra Commission, even though he greatly regretted having yet another assignment that would take him away from his family for long periods of time.

Pellagra was considered an intractable puzzle. Known and feared around the world, it first appeared in the United States during the Civil War. By 1908 about a hundred thousand Americans were victimized every year, most of them poor rural sharecroppers. The disease was especially prevalent in confined situations, such as orphanages, prisons, and asylums.

Its name derived from the Italian words for "rough skin," pellagra typically begins with reddish, hardened flesh, often producing a telltale butterfly rash on the nose. This is only the first stage of the so-called disease of the four Ds: dermatitis, diarrhea, dementia, and death. In the second stage diarrhea is accompanied by ulcers in the sufferer's bowels and mouth. In the third, victims grow confused, begin to hallucinate, and eventually go mad. Pellagra victims have historically been treated like lepers. The disease had a 50 percent mortality rate.

Public reactions of the early twentieth century are familiar from the bubonic plague experience. To some extent, Italian immigrants were blamed for carrying pellagra; they were conveniently a distinct foreign group with identifiable cultural characteristics. But former slaves were also considered likely carriers, as were insects. Even the heat and sunshine of the southern spring were proposed as causes, but the two most widely credited theories put the blame on bacteria or the eating of "bad" corn.

Medical researchers assumed that like the plague and other bacterial infections, pellagra must be caused by a microbe, and some scientists even isolated "pellagra bacteria." Others, after pondering the anecdotal evidence from institutions of confinement, assumed that the disease was contagious; its microbe must thrive upon and be spread by unsanitary living conditions.

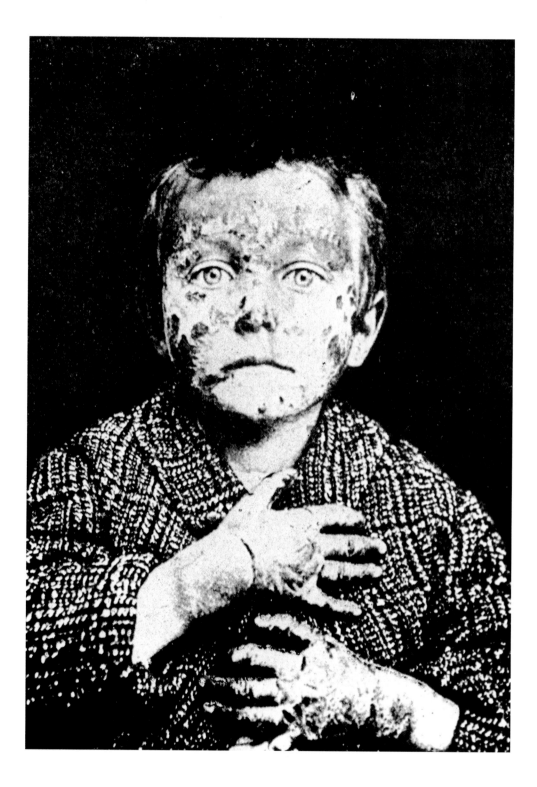

These conventional and proper assumptions were contradicted by Goldberger with the simplest of expedients: In 1915 he went to where the disease was most virulent.

In the Georgia State Sanitarium for the Insane several hundred inmates suffered from pellagra, but Goldberger discovered a startling fact. Not a single case existed among the 293 nurses, attendants, and other employees in daily contact with patients. Convinced that no germ honors status boundaries, he caught on in a flash of insight: "This peculiar exemption or immunity was inexplicable on the assumption that pellagra is communicable."

At other southeastern institutions he found the same phenomenon. With growing excitement he began to suspect that some essential difference between the living conditions of keeper and kept could explain the origins of the disease. In *The Butterfly Caste: A Social History of Pellagra in the South,* Elizabeth Etheridge re-creates the young scientist's splendidly simple but unprecedented insight: "He observed that the inmates ate the monotonous 3-M diet of the poor South: meat, mostly fatback, and meal, mostly corn meal, and molasses. In contrast, the attendants had access to fresh meat, milk, eggs and vegetables."

Probably aware of the discovery of vitamins by Frederick Hopkins and Casimir Funk, who showed that beriberi is caused by dietary deficiencies, Goldberger realized that some sort of nutrition problem might be the activating agent in the pellagra epidemic.

"Beans and Beans!!!"

Goldberger set up an experiment. Since the disease did not occur in animals, he necessarily used human subjects.

At a cost of seven hundred dollars a month per institution, the state of Georgia agreed to fund a study involving pellagra victims in two orphanages. The children's glum diet would be supplemented with fresh meat, milk, and dried legumes. At the same time Goldberger ordered that nothing be done about the unsanitary conditions that prevailed. Thus diet became the one variable in his experiment.

Overnight he was a pied piper of the orphanages. At first the sick children clam-

---

It is not difficult to understand why the disfiguring symptoms of pellagra caused it to be regarded by most as a disease caused by bacteria. In fact, Joseph Goldberger's discovery of the actual cause produced a revolution in the understanding of the connection between diet, socioeconomic conditions, and certain illnesses. Today, medical science lists about fifty compounds that are considered essential nutrients in the healthy human diet.

ored to thank him for the good food, for reasons that need no explanation, and then the miracle happened, for within a few weeks the symptoms of pellagra began melting away. Color returned to the cheeks of orphans who had been listing toward death, and previously healthy children remained healthy.

An exuberant Goldberger wrote the news to his wife: "It seems possible that pellagra can be wiped out in our South by simply getting people to eat beans. Beans. BEANS and BEANS!!! If they will eat enough, they can eat anything else they D[amn] please."

In fact, as he soon recognized, beans played no role in the dramatic recoveries. Meat, milk, and fresh vegetables had effected the cure. Goldberger published his findings in an article, "The Etiology of Pellagra," but to his bewilderment few people agreed with his dietary reasoning. Other scientists were comfortable with the more conven-

tional theory of the time: The health workers did not succumb to the supposed pellagra germ because they were stronger and healthier than inmates, who were by definition weak and degenerate. The experiment with the orphans' diet was resented as sly criticism of the traditional southern way of life. End of discussion. In other words, rash Goldberger had failed to spot the pertinent bacterium and was impolitely defying social convention. "Impolite" he certainly was, for he frankly accused southern employers of paying their workers so poorly and feeding them so meagerly that they and their families fell ill.

Goldberger became even more determined to nail down the proof scientifically by taking the reverse approach—i.e., using a vitamin-deficient diet to induce pellagra in healthy people. Aided by the governor of Mississippi, who promised full pardons to any twelve healthy convicts who would volunteer, Goldberger assembled a group known as the pellagra squad. While the other prisoners ate standard prison fare, which was hardly a contemporary nutritionist's dream but included vegetables from the prison farm, the members of the squad would be fed the typical 3-M southern diet and denied fresh milk, lean meats, and all vegetables. They were given clean clothes and isolated in special buildings that were screened to keep out insects and scrubbed rigorously each week to deter any microbial infection.

The prisoners gradually learned that volunteering within an institutional framework is rarely a good idea. After a mere three weeks they began to complain of lost weight, steady pain in the back and sides, and mouth sores. But the mental changes were even more disturbing and dramatic. The volunteers became confused, moody, and alarmed. Two wrote to the governor begging to be excused. A lifetime of hard labor, they wrote, would be better than one more week of this pain and disorientation. Later one was quoted in the press: "I have been through a thousand hells."

After five months of this slow torture the telltale pellagra rash bloomed out, overjoying Goldberger and his associate, Dr. George Wheeler. Outside physicians confirmed that seven of the squad members, or more than 50 percent, had contracted the

It was not unusual for the typical meal of a rural American family early in the twentieth century to be dangerously deficient in essential vitamins, especially during the 1930–1932 drought afflicting Arkansas, West Virginia, and Virginia, when Lewis Hine took this photo. Goldberger did not learn exactly what substance his pellagra patients lacked, although he saw that certain foods provided it. Other researchers discovered it to be niacin, supplied in red meat, poultry, and liver.

disease. This was a rate about three times what Goldberger felt he could reasonably hope for. The prisoners were immediately pardoned. More important, they were instructed how to eat in order to regain their health.

Twenty Wasted Years

Unfortunately, these two sensational experiments—the reviving of the afflicted orphans and the horrible debilitation of otherwise healthy adult prisoners—met with hardening skepticism. This was all the more astonishing since here was a noted infectious disease specialist in the bacteriological age who had demonstrated fairly unusual theoretical flexibility in finding a solution outside his specialty.

To be fair, Goldberger did not exactly encourage a dialogue. "Let the heathens rage!" he wrote, analyzing his critics as "blind, selfish, jealous, prejudiced asses."

With a characteristic combination of showmanship and self-confidence, Goldberger began holding "filth parties" to disprove the bacterial theory. The first involved only himself and Wheeler. Each had about a fifth of an ounce of blood drawn

from a pellagra sufferer injected into his left shoulder. Then, enlivening the spree, they took turns rubbing secretions taken from the patient's nose and mouth into each other's nose and mouth. Nothing happened.

Three days later Goldberger swallowed capsules filled with the urine, feces, and skin of several patients severely afflicted with pellagra. He suffered diarrhea for a week, but no symptoms of incipient pellagra. In a final "filth party" his wife and five other volunteers downed the same kinds of capsules.

Left: *In this romanticized painting, Goldberger* (center) *muses on a miraculous transformation at the Baptist Orphanage in Jackson, Mississippi, in 1914. Following his addition of fresh meat, eggs, and milk to the institutional diet, formerly ill orphans glow with newfound health and good cheer. The reality was just as dramatic: A balanced diet "cured" pellagra. Below: Goldberger* (seated) *and Dr. George Wheeler pose with the prisoners who took part in the pellagra squad. Not long after they were restricted to the kind of diet usual in many poor homes in the South, they began to suffer so horribly from physical pain and disorienting hallucinations that many begged to be sent back to "hard time."*

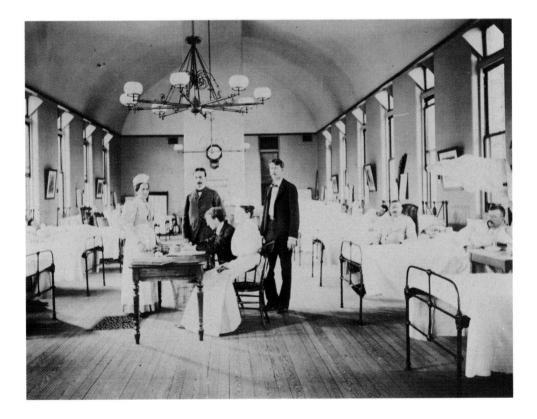

"If anyone can get pellagra [this] way," he declared on June 25, 1914, "we should certainly have it, good and hard." None of them did. That should have proved to the skeptics, if not the squeamish, that pellagra was not caused by a germ and could not be transmitted from its victims to healthy persons.

But, to Joseph Goldberger's lifelong frustration and amazement, neither established medicine nor government agencies would take him seriously. At his death from cancer in 1929, three hundred thousand people in the United States were suffering from pellagra.

Only five years later, scientists finally spotted the missing nutritional essential, a B vitamin called niacin. Why had this dietary deficiency been limited to certain social classes in one area of the country? In the main, the answer lay in the thin pay packets of the working poor who labored in cotton mills or farmed cotton as sharecroppers. There was another curious factor. Decades before, millers had decided to produce fine-grain meal rather than the traditional whole-grain meal. That process had inadvertently removed niacin, thus aggravating one of the most devastating epidemics in U.S. history.

At last the federal government responded by ordering that milk and some other foods be fortified with niacin. Pellagra disappeared forthwith from the American scene. But Goldberger should be remembered, as some defensive southerners quite accurately recognized, for the social dimensions of his work. As the wider world came to understand only toward the close of the twentieth century, he saw that poverty can play a critical role in disease—almost any disease in any country. Just as important, he stressed the role that the individual—if respected enough to be educated about a disease—can play in his or her health care. Pellagra to the side, how many millions have suffered or died because others did not recognize then or now the validity of Goldberger's insights?

Even as the potential impact of Goldberger's work was being ignored, researchers were discovering other connections between environment and health. It was known that people never exposed to sunlight, and therefore deficient in vitamin D, would get rickets, a bone disease that results in bowlegs and other deformities. A diet lacking in vitamin C would cause scurvy.

In other words, these researchers began to visualize the human body as a kind of bipedal chemical processing plant. When inappropriate materials slipped onto the assembly lines or something went wrong in the conversion of materials, there could be serious malfunctions.

This was a surprising new thought. Disease—far from inevitably resulting from assaults by germs or other foreign organisms—could be caused by an enemy lodged within.

"Melting of the Flesh"

Since it was first identified more than two thousand years ago, diabetes mellitus has been one of the most horrendous and inexorable of such internal foes. Physicians of ancient Egypt recognized by 1550 B.C. that the need to urinate constantly was an infallible symptom. The ancient Greek physician Aretaeus, writing in the second century B.C., poetically and accurately described the final stages of a diabeteslike disease: "[T]he melting of the flesh is rapid, the death speedy." He named the disease "diabetes" from a Greek word for "siphon" because of the frequent urination.

The stark contrast between a diabetic child's condition before and after insulin treatment is hauntingly evident in this pair of photographs taken in the 1920s. The breathtaking recovery occurred after only four months of insulin therapy. These and similar images were printed along with the article in the Journal of the American Medical Association *that first explained insulin and its effects to physicians.*

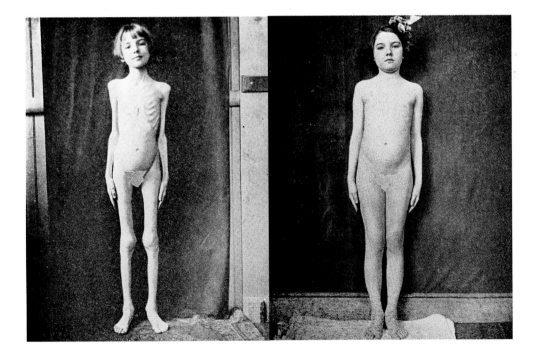

Early in the twentieth century medical science was aware that diabetes was caused by the body's inability to metabolize all of its required food, especially carbohydrates. Instead, most of these essential nutrients pass through unused, and the victim becomes dehydrated and begins to lose weight. As a side effect the level of sugar in the blood rises very high. Racked by excruciating thirst and hunger, diabetics slowly waste away from lack of nourishment.

There was no cure, and the chief therapy—i.e., feeding patients less since they were unable to process much food—prolonged the cruel agonies. Carrying this idea to the extreme, Dr. Frederick Allen, a specialist in diabetes treatment at the Rockefeller Institute in New York City, enjoined his patients to lose 30 percent of their total body fat. This treatment was deservedly known as starvation therapy. His patients, looking like concentration camp victims in midcentury photographs, agonizingly died of malnutrition or the disease.

Rarely has there been such a compelling example of the cure's being worse than the disease, but this horrible regimen could buy a further year or two of life. The consequences have been memorably described by Michael Bliss in *The Discovery of Insulin:* "There are terrible stories of diabetic patients stealing hot food from ovens, burning their hands to do it, eating toothpaste and bird seed left in their rooms. And even if they did manage to comply, it was heartbreaking to see them shrink away to nothing but skin and bones and then fall into coma. And once they went into coma there was no hope."

One of Allen's most prominent young patients was Elizabeth Hughes, the eleven-year-old daughter of Supreme Court Justice (and later Republican presidential nominee) Charles Evans Hughes. Weighing seventy-five pounds when the disease struck, Elizabeth was brought down to fifty-five pounds by Allen's combination of fasting and a low-calorie diet of four to six hundred calories a day. Every gram of food was carefully weighed. Meals included bran husks stuffed with sawdust to provide bulk without calories. Vegetables were boiled three times over to leach all caloric content out of them. Of course, the diagnosis itself was a death sentence, though grimly protracted.

Allen was so convinced of the soundness of his treatment that he believed diabetic children should be locked in their rooms to prevent them from getting their hands on food.

The Mystery of the Pancreas

Since 1889, researchers knew that diabetes was somehow linked with the pancreas, a yellowish gland attached to the small intestine beneath the stomach. The pancreas facilitates digestion by pouring juices into the stomach; it also breaks down carbohydrates and sugar by sending a different secretion into the bloodstream. Diabetics were known to have high levels of sugar in their blood. Experiments had shown that a lab animal would contract diabetes when its pancreas was removed.

The implications for cure seemed obvious: Isolate the secretion created for the bloodstream, give it to diabetics, their bodies would begin metabolizing carbohydrates again, their levels of blood sugar would decline, and the disease would be slowed or conquered. Unfortunately attempts to alleviate the illness with pancreas extracts were more likely to kill patients, since the extracts were not sufficiently pure.

In 1920 a very unlikely candidate decided to solve the riddle that was engaging highly regarded scientists in some of the world's most prestigious research laboratories. Frederick Banting, a young Canadian surgeon unable to attract enough patients to support himself, happened to read an article about the pancreas. He later recalled the immediate aftereffects of this fateful browse: "It was one of those nights when I was disturbed and could not sleep. I thought about my miseries and how I would like to get out of debt and away from worry. Finally, at about two in the morning, the article had been chasing through my mind and the idea occurred to me that one might be able to obtain this internal secretion." It was, of course, a revelation that was pretty much shopworn, but Dr. Banting's ignorance or naïveté produced one of history's few medical miracles.

A painfully shy, virtually inarticulate young man, he had not distinguished himself in his academic preparation and even misspelled "diabetes" in his notebooks. Yet with astonishing crust he immediately sought help from J.J.R. Macleod, a University of Toronto professor of physiology who was an internationally respected expert on the metabolism of carbohydrates. Understandably dubious, Macleod reluctantly agreed to give Banting a few months' access to a cramped and dingy lab, ten dogs to be used as experimental animals, and the services of a research assistant, Charles Best. Banting himself would be paid nothing. Work began in the summer of 1921.

Banting's game plan had the textbook experimental virtue of simplicity: Operate on the dogs to block the ducts that produce the pancreas's stomach secretions, let them shrivel to nothing, then use the section of the pancreas that remains—that is, the part that supposedly produces nothing but the bloodstream secretions—to produce a curative extract. The cells that manufacture these secretions are known as the islets of Langerhans.

"Sugar Free!"

The initial results were disastrous. In ghastly working conditions exacerbated by oppressive heat, the experiments uniformly failed. The original ten lab dogs died, and the two men, like inspired ghouls, scoured the Toronto streets by night to pick up ill-fated strays. One by one they too died.

Barely stopping for sleep, since the meter was running on the university lab, Banting and Best fumbled through their operations for six or seven weeks. Then, with a rush of exhausted relief, they finally obtained the extract they were seeking. They injected it into dogs suffering severe diabetes because their pancreases had been removed. When the blood sugar in these dogs was measured, the levels had plummeted. Ecstatic, Banting underlined the words "sugar free!" in his lab notebooks.

But at summer's end the more experienced, cautious Macleod did not share the exhilaration of the two unknown, unconventional young men. "Of course," he said coolly, "one result is no result."

Macleod was bureaucrat as well as scientist. He did not believe, as Banting did, that the project should take priority over all other work in the university laboratories.

Their relationship eventually went ballistic. Banting was ready to repeat and refine his experiments for Macleod, allowing for all possible variables according to basic principles of scientific research, but he demanded better facilities, more dogs, more assistants—and a salary, since he was now broke. Macleod was willing only to extend, not to beef up, his commitment to Banting.

Still, despite heated arguments and scorching language, these innately antagonistic men agreed on research goals. At Macleod's suggestion, Banting decided to find out how long a diabetic dog could survive on a regimen of the pancreas extract. To obtain enough of the substance, he used ground-up calf organs, readily available in sizable numbers at any slaughterhouse.

It took until November 20 to get everything set in place. Two diabetic dogs, No. 27 and No. 33 (also known as Marjorie), were each given two daily injections of slaughterhouse extract. On December 2, No. 27 suddenly died, but after more than two months Marjorie was not only alive but apparently thriving. Ironically, the chopped-up calf organs were just as effective as the material produced by months of work tying off pancreatic ducts.

At Macleod's suggestion, biochemist J. B. Collip joined the team. Right away his expertise produced extracts freer of toxic contaminants than any made so far by the enthusiastic amateurs, and the results with animals also improved.

As early as January 8, 1922, Collip could note down one of the century's most significant medical discoveries. He did so with professional caution: "We have obtained from the pancreas of animals a *mysterious something* which when injected into totally diabetic dogs completely removes all the cardinal symptoms of the disease. If the substance works on the human, it will be a great boon to medicine."

The "mysterious substance," the pure form of the pancreatic hormone, was named insulin by the team. Three days later at the Toronto General Hospital Banting in-

Far left: *J.J.R. Macleod, a fastidious administrator, was never comfortable with Frederick Banting but did recognize that the younger physician might have the key to producing an injection from pancreatic materials that could alleviate diabetes.* Left: *Charles H. Best* (at left in photo) *and Frederick Banting pose with one of the first dogs to be restored to health with insulin after diabetes was surgically induced in the animal. This breezy photo taken atop the University of Toronto's medical building belies the steamy, cramped conditions in the lab where the two men learned to produce the life-prolonging extract. The actual lab, now unrealistically neat and clean, can be seen today at Toronto's Museum of Technology.*

jected some of his own extract into a fourteen-year-old diabetic boy, Leonard Thompson, who at sixty-five pounds was on the brink of death. Not only did the treatment fail to lower his blood sugar levels to any meaningful degree, it also caused a dangerous reaction. Banting's extract contained toxic contaminants.

Almost two weeks later, on January 23, young Thompson was given an injection of Collip's differently obtained extract. A urine test revealed that his blood sugar dropped. He was already becoming more active, looking better, and reporting that he felt stronger. Like Marjorie, he had just entered medical history; this was the first successful trial of insulin on a human diabetic.

"Immortality, Glory and Dollars"

This astonishing breakthrough had an unexpected result, a dramatic breakdown in the team effort.

"I've done it," Collip crowed to Banting and Best. "I know how to do it."

"How did you do it?" Banting sensibly asked.

"I'm not going to tell you."

Banting leaped upon the smirking biochemist, nearly throttling him, but things were going to get even worse all around.

Historian Bliss has summed up what was beginning to happen in a lab whose efforts had defeated one of the most obstinate of the ancient diseases created from within: "They knew that they had a discovery that was going to save millions of lives. Somebody had also talked about patenting the discovery. As often in science, team work and cooperation are all very well until immortality, glory and dollars start to intrude."

Worse yet, now that Collip was pressured to turn out large quantities of the extract, he unaccountably failed to reproduce the formula.

Nonetheless, word of the apparent cure spread rapidly, and diabetics and their families, desperate for a share of the extract, besieged the university lab. Nothing was forthcoming. Again and again the team could not produce pure insulin. Leonard Thompson was sent home in the spring, but without any extract to maintain his improved health. According to the scanty records existing at the Toronto General Hospital, it seems that some patients, having grown strong on the initial supply of insulin, were returned to their starvation diets. One of them, a young girl, slipped gradually into a coma.

Fortunately for diabetics, however, the U.S. pharmaceutical company Eli Lilly learned to produce purified insulin on a massive scale, and Dr. Banting and others could at last save the afflicted. This, after all, had been the original goal of their contentious science.

Elizabeth Hughes, whose father's eminence could work no miracles, appeared at the Toronto General Hospital in mid-August. A five-foot-tall sixteen-year-old who had dwindled to forty-five pounds, she was only days from death. With pure insulin, she became Banting's prize patient, responding so favorably that she began gaining weight by the day.

"This stuff is unspeakably wonderful," she wrote to her mother.

Professional reviews were no less exuberant. Dr. Elliott Joslin of Boston, a prominent diabetes specialist, called insulin "one of the greatest discoveries of the century, the beginning of a new epoch in medicine." He quoted the Old Testament Book of Ezekiel

A biochemist with the skills necessary for refining a reliable version of insulin with as few side effects as possible, J. B. Collip was the fourth member of the quarrelsome team whose achievements have improved and prolonged the lives of millions in this century. Their extract came from calves, but the insulin manufactured in large quantities by Eli Lilly and Connaught was derived from pig pancreas.

to describe the effect of treatment: "I will lay sinews on you, and will cause flesh to come upon you, and cover you with skin, and put breath in you, and you shall live."

"The Medical Hotel"

After a few years, it was clear that insulin did not "cure" a disease; it replaced a missing substance in the body. That was a tricky business, as diabetics discovered. If they took insulin and tried to return to a normal diet, they risked death.

Joslin was among the first physicians to learn the true state of affairs; in effect, insulin had transformed diabetes from fatal disease to chronic ailment. To prevent life-threatening complications, it would be necessary to exert continuing watchful control over the changed course of the disease.

Dr. Joslin became a missionary for informed insulin usage, setting up a so-called medical hotel where diabetics lived for a while, learning how to prepare the right kinds of food, prick their fingers to test for blood sugar levels, exercise, and inject themselves with insulin. Each patient had to learn how to regulate insulin intake, how to maintain a proper dietary balance. For example, up to 60 percent of calories should come from carbohydrates. In the strangely moralistic atmosphere of this institution, patients who did not thrive were considered at fault. Patients who followed the rules would flourish. Himself an ascetic of a sort, Joslin once called his staff into conference when a case of rare Scotch whisky arrived from an admirer. Lecturing with fervor upon the evils of strong drink, he poured the contents of each bottle down the drain.

But neither enthusiastic advocacy nor rigid discipline could disguise the frustrating clinical reality: Insulin could only prolong the diabetic's life. It could not prevent such painful and disabling complications of the disease as kidney failure, blindness, and loss of limbs. Patients with early-onset diabetes could find their life expectancy increased twenty-five–fold. But Leonard Thompson, who cheated death with mere hours to spare, died from pneumonia at age twenty-nine. His life was both doubled and profoundly foreshortened by the imperfect chemistry of his own body, and also by his inability to follow a strict dietary regime closely.

Elizabeth Hughes, shown here with her mother and her father, Supreme Court Justice Chief Charles Evans Hughes, was one of the first diabetic children to be brought back from the brink of death by insulin therapy. Gradually, it became clear that the hormone would be effective only in combination with a carefully controlled diet and that some diabetes-related afflictions would eventually develop.

Still, the discovery of insulin was a meteoric achievement, and there was no contest, in the minds of the panel awarding the Nobel Prize in physiology or medicine in 1923. Banting and Macleod won.

The rest of the story was as inevitable as it was depressing. Banting, the outsider who thought of himself as fighting for his inspired idea against entrenched academic forces, was furious that the coolly dismissive Macleod, not his fellow outsider Best, had been chosen to share the glory. Banting split his prize money with his assistant and refused to travel to Sweden to accept the Nobel. For his part Macleod shared his money with Collip and did not hesitate to cross the Atlantic for the awards ceremony.

At the formal dinner in Stockholm, a scientist happily outside the fray said, "In insulin there's glory enough for all of them." But for the next sixty years, until all the major players died, neither of the two Canadian Nobel laureates would speak to each other, and there was continuing gossip and contention about the actual roles played by all four men in the isolation of insulin. Bliss has sorted it out this way: "The key to the breakthrough was the work of the senior scientists Macleod and Collip, who were able to take the work of inexperienced but enthusiastic researchers and refine it."

Such a reasonable division of glory was impossible for Banting and Best, Macleod and Collip. Even as they went on to other projects and received still more awards and grants and high positions, they were tormented by ill will engendered by the major achievement of their lives. The odyssey of science is a journey taken by human beings.

The fight against diabetes continues. In the United States at the end of the twentieth century, according to the American Diabetes Association, something like sixteen million people—that is, 6 percent of the population—suffer from the disease. Despite the miracle of insulin therapy, perhaps one in two may either have to have a limb amputated because of diabetes-related gangrene, lose the use of his or her kidneys and need a transplant, suffer serious nerve damage, or go blind from diabetes retinopathy.

Finding Magic Bullets

Today's contemporary assumptions about magic bullets—remedies that effectively combat or destroy a specific bacterial disease or infection—had little foundation as late as the fourth decade of the century. Scientists were learning a great deal about the nature of certain diseases, virtually nothing about how to cure them. For example, millions of people around the world died from influenza in the early twentieth century, including a million in the United States in 1918 alone.

The first magic bullet, so named by its amazed discoverer, was a cure for syphilis. To treat this sexually transmitted disease, which was widely prevalent one hundred years ago (and is still one of the seven most common infectious diseases), the German chemist Paul Ehrlich tested 605 compounds over a three-year period. In 1909 he finally found success in killing syphilis bacteria with No. 606, which he developed as the drug salvarsan, the first drug that could actually cure a disease.

There was a slight problem. Because the cure's active ingredient was arsenic, the famously lethal poison, dosages of salvarsan had to be carefully measured and cautiously administered. After several patients died, rudimentary drug trials were set up to teach physicians how to handle this effective, if tricky, medicine.

In the 1920s an obscure researcher working in the inoculation department at London's St. Mary's Hospital became the most proficient instructor in salvarsan trials. In fact Alexander Fleming earned the nickname Private 606 for he was unfailingly discreet about the embarrassing illness when he treated various proper members of Edwardian society.

Fleming's chief interest, however, was bacterial infection, particularly in skin and surface wounds. During World War I he had served in France with the Royal Army Medical Corps and necessarily learned about the effects of various kinds of infection— gangrene, streptococcus, staphylococcus—on the wounded soldiers brought in from the killing fields. These battlegrounds were alive with microorganisms because they were usually level farmlands that had been richly fertilized. Harsh chemical antiseptics were used to treat surface infections, but they destroyed the leukocytes that provide the body's natural defense against bacteria; such therapy could encourage rather than cure infections. Amputation was often the final recourse when limbs became infected, as it had been half a century before in the American Civil War, but it worked. For internal infections there was no treatment, and death was the probable outcome. It is thought that more young warriors died from bacterial infection than from bullets or mortar shells.

At St. Mary's after the war Fleming sought a way of defeating infectious bacteria without either compromising the body's natural defense system or destroying the affected tissues. His work eventually changed the world, but not overnight. It is a story so odd that once his name became virtually a household word, it inspired a persistent but completely fanciful legend.

In this nicely rounded tale Fleming forgets to close a lab window before taking his annual holiday in Scotland in July 1928. Coincidentally the lid on one of his culture plates of staphylococcus bacteria is somehow jarred loose. Even more coincidentally, while he's away, an airborne mold is wafted into the room and lands upon the bacteria colony. When Fleming returns, he sees a remarkable sight: There is a clear space between the edge of this green mold and the staphylococcus colony. Obviously the mold has secreted something that kills bacteria.

This is a wonderful story for those who prefer to believe that some of the greatest discoveries of humankind are completely serendipitous, a kind of lottery only a little swayed by tenacity and talent and sharp thinking.

Yet chance and character traits did figure strongly in the true story. The mold, a rare strain apparently being studied by researchers on the floor below Fleming's lab, probably flew up the stairwell. It landed on an open dish because Fleming was something of a slob, forever leaving things lying about. Finally, a shift in temperature ensured that the mold survived and the bacteria grew. Still, none of this would have made any difference to the average life expectancy of the human race if the man himself had not glanced at the contaminated plate before washing it out and instantly understood what he saw.

If not as neatly ordered as the mythological version, this process of scientific discovery was perhaps more extraordinary, as was the activity of the sepsis in the mold, the *Penicillium notatum* that Fleming dubbed "penicillin." This substance killed a bacterium by attacking a certain part of its cell wall. It was hugely effective, but that gave Fleming pause. Would it also pulverize animal or human cells?

He injected penicillin in bacterially infected lab mice with a promising result: The bacteria died; the mice prospered. Unfortunately Fleming then had difficulty isolating any sizable quantity of his discovery. In addition, the penicillin showed no effects upon some types of bacterium. He may have performed limited experiments on humans, but he was most interested in the potential for dealing with his main passion, and the results for surface infections were disappointing. The substance was nontoxic, but its

Horrified that more soldiers in World War I were killed by infections than by gunfire, Alexander Fleming determined to find a stable, nontoxic antibiotic. He discovered lysozyme, an antimicrobial enzyme found in human nasal secretions, in 1922 but was unable to isolate it in sufficiently concentrated form to combat infection. Six years later, in an accident that became science only because he recognized what it meant, Fleming discovered the very antibiotic he sought.

beneficent effects seemed to diminish quickly. *Penicillium notatum,* as a magic bullet, was proving to be scattershot.

Fleming wrote about penicillin in the *British Journal of Experimental Pathology* on May 29, 1929, but left out the very loud bells and whistles: that it wiped out systemic infections in mice. He was even tentative about its proved effectiveness in killing bacteria at all. He set the most important work of his life to the side and turned to other pursuits.

With hindsight, it is easy to criticize Fleming as being blinded by stubborn obses-

sion, but his evidence was not compelling. Besides he was surely recalling the long rows of young soldiers dying from infections, not wounds. He yearned to find a topical anti-septic. Penicillin did not perform this function well. While his experience with the in-fected mice could have suggested that he had found something much more important—the first antibacterial since salvarsan—there was another factor in the air: No one in the scientific community believed that a substance like penicillin would ever be found.

As penicillin was being shelved, the German chemist Gerhard Domagk was look-ing for antibacterial compounds in the 1930s at the giant I. G. Farben chemical company in Frankfurt. His method was to explore the effects of certain dyes on streptococcus bacteria. Whenever a dye stained a bacterium, he reasoned, it must somehow be reacting with the microbe, sticking to it and changing it. Domagk sought a way to use dye as a messenger to deliver a substance that could kill or weaken the germ.

Success came when a red dye containing sulfonamides cured a streptococcal bacterial infection in mice. This result was more or less what Fleming had seen in sim-

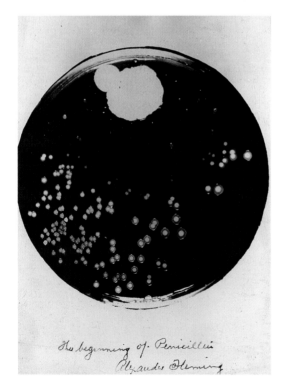

One of the most famous images in medical science, Fleming's own photo of the culture plate that first revealed penicillin's effects shows the mold as a large white area. The bacteria, known as lysis-staphylococci, have been wiped out nearby, surviving only at the far rim of the plate. It would be another fifteen years before other researchers found a way to manufacture a practical drug from the remarkable mold.

ilar circumstances with penicillin, but neither scientist was aware of the other's work.

A few patients with very severe or critical strep infections were given Domagk's drug. One was Franklin Delano Roosevelt, Jr., whose infection seemed to vanish overnight. The sulfonamides work by preventing microbes from multiplying, not by killing them, thus giving the body's defense system enough time to come on-line and rout the invaders. The so-called sulfa drugs became an international sensation.

They were standard issue on both sides in World War II. If shot or injured by a mine, soldiers were instructed to sprinkle their packets of sulfa powder directly on the wound. For systemic infections, sulfa injections were available at field hospitals.

Penicillium Redux

Sulfa drugs are not effective with all types of bacterium, and they can have unpleasant side effects, even severe reactions for many patients. For these reasons, two scientists at Oxford University—Australian-born pathologist Howard Florey and Ernst Chain, a Jewish refugee from Nazi Germany—were trying to find a better antibacterial, even as World War II raged on the Continent. Piqued by Fleming's slight and tenuous article on the effects of *Penicillium notatum,* they rang him up in 1939. He had kept the tantalizing mold. They took a portion and, after about a year and a half of intense work, were able to produce a mere 0.0035 ounce of pure penicillin.

On May 25, 1940, they tested it on half of a group of mice infected with streptococcus bacteria. The untreated mice died within sixteen and a half hours of being infected; those given penicillin thrived. Chain and Florey knew they had found something major. Because the outcome of the war was very much in doubt, they rubbed spores of the mold on their clothing, planning to carry penicillin to America if Britain was invaded.

Then they looked for patients so desperately ill that penicillin would be their last hope. Six subjects were chosen, including a policeman who had cut himself on a rose thorn and was dying from staph infection. Penicillin improved his condition, but he relapsed when the penicillin supply ran out. Desperate both to save the man and to probe the virtues of this encouraging substance, Chain and Florey even recovered traces of penicillin from his urine, but it was not enough. The policeman died. Yet the other five patients were cured.

The next obstacle was that the *Penicillium* mold does not naturally produce very much penicillin, and human patients require about three thousand times as much as an infected lab mouse. It had taken nineteen researchers working around the clock to produce enough for the six patients. Norman Heatley, the sole surviving member of the Oxford research team, hit upon the idea of using bedpans to grow mold. "Even with hundreds of pans of growing mold," he recalls, "we could retrieve only a small amount of the drug. . . . But it was enough to test it on our next patient, a boy with a small body, who needed less of the drug."

The frail child was in danger of going blind from a serious eye infection. The bedpan expedient had produced enough penicillin for a week's worth of injections. It was enough, as it turned out, to cure him completely. With this joyous result the team became even more highly motivated to produce penicillin at a faster rate.

Eventually two factors came beautifully together. In Peoria, Illinois, Florey and Heatley discovered the Northern Regional Research Laboratory of the U.S. Department of Agriculture, a facility that had learned how to grow molds in large quantities. After several months' work scientists there developed a more fertile growth medium for the mold from corn syrup. Then, by chance, they also found a previously unknown strain of

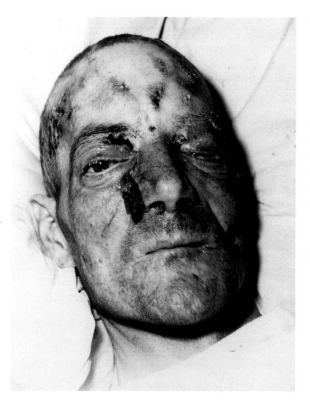

From the scratch of a rose thorn, a British policeman developed a life-threatening staph infection. A team of scientists at Oxford University's Dunn School of Pathology worked frantically to purify penicillin. The drug worked, but the patient died when supplies ran out. An American pharmacologist, A. N. Richards, helped convince U.S. drug companies to invest in developing the antibiotic, which was extraordinarily successful against severe infections in World War II and sparked research efforts around the world to find other antibiotics.

Penicillium growing on a rotting cantaloupe; it was much more prolific than Fleming's original mold.

The next step was unquestionably one of the most triumphant cooperative ventures in medical history. In the interest of mobilization for war in Europe, America's pharmaceutical companies pooled their resources. From January through May 1943 they were able to produce four hundred million units of penicillin. By the end of the year monthly production was some nine billion units. The nation's military aims and vast industrial resources were key to the success of penicillin.

For the first time there really was a miracle antibacterial drug in the hands of treating physicians. It played an enormous role in the Allied victory; millions of soldiers survived wounds that would have killed them only the year before. In the future hundreds of millions more lives were to be saved by penicillin.

The ability to make use of penicillin in fighting bacterial infection marks the beginning of the era of exuberant optimism. To a degree that is almost impossible to comprehend, penicillin radically changed the composition of the human family. By the end of our century millions of people could look back only one or two generations and find a direct ancestor who would have certainly died without penicillin treatment.

The Forbidden Organ

Ironically, not only did the trauma and chaos of World War II help affirm the lifesaving properties of penicillin and sulfa, which had been developed in pressured if subdued laboratories, but also in exposed field hospitals, on the battlefields, indeed everywhere connected with medicine in all the major conflicts of the century, much was learned about both surgical techniques and the prevention of disease. Even before World War I surgeons had achieved a high success rate with various kinds of operations on the human body—not just the removal of tonsils or the appendix but also complex procedures involving the brain or the blood vessels of the vascular system.

Historically, as former Harvard University professor of surgery Francis Moore has explained, the heart was considered the forbidden organ: "You stop the heart, you kill the patient. That's that. Early attempts just to open up the chest were disastrous. The patient's lungs would immediately collapse."

The medical community had concentrated energetically on fatal infectious diseases, then the leading causes of death, while heart problems had received comparatively little treatment by the first decade of the twentieth century, primarily because surgeons sensibly feared opening up the chest. Occasional attempts to operate on diseased or malfunctioning hearts had rarely been successful. It was as late as 1910 before James Herrick, an internist, wrote the very first article describing a heart attack for an American professional journal. In the same year Swiss surgeon Ferdinand Sauerbruch made a considerable contribution to the technology of heart surgery. To prevent a patient's lungs from collapsing when the chest was opened, he designed an operating chamber that reduced air pressure around the chest area.

But it was the Great War that forced surgeons to penetrate the forbidden organ, simply because with certain types of wounds there was no other choice. U.S. Army surgeon Dwight Harken found himself dealing with young men who had been evacuated from the theater of war with bullets or shell fragments still lodged within their beating hearts. Whether the foreign object had come to rest in muscle or floated freely in a chamber of the heart, it could not be allowed to remain; the slightest jar might lacerate or block the heart, causing death. At the time, however, any effort to remove such intruders surgically was considered equally likely to be fatal. Few surgeons would take the risk.

Harken was not numbered among them. After trials with lab animals, he developed a technique for cutting into and cleaning out a living heart. He limited his first human trials to fourteen badly wounded soldiers. He cut a small hole in the wall of the heart, then carefully stitched it open. Next, guided only by his sense of touch, he probed for the metal fragments and pulled them free and out of the body. Nothing like this cardiac operation had ever been attempted before, yet all fourteen young men were back on their feet within days. Conventional medical assumptions were disproved; the legendarily delicate heart was in fact one tough hunk of muscle. Harken eventually performed 134 of these operations without a single fatality.

But although he did intrude into the pump of life, this was still "closed-heart surgery." Three years after the war Harken—and coincidentally a Philadelphia surgeon named Charlie Bailey—used a new technique to try to correct mitral stenosis, a condition in which an essential heart valve is so narrow that it cannot open properly. Again, after cutting a small entrance hole, both surgeons probed blindly with their fingers, manually widening the valve to normal size.

Bailey's first five patients died, but the sixth survived for twenty-three years. The operation became a widely accepted technique. But this simple procedure was no answer for patients suffering from more complicated heart defects. To address many of these diseases or malformations, the surgeon would have to cut into the heart and open it up in order to analyze visually, then operate on the specific problem. Fine in theory, but one obstacle seemed insurmountable: The heart would have to be stopped from beating and drained of blood. This would cut off essential oxygen to the brain, which would suffer irreversible damage within two to four minutes.

"This Is It"

Bill Bigelow, a young Canadian surgeon, reflected upon this difficulty in the early 1950s. In that decade, medical science became less concerned with infectious diseases and more interested in such chronic and functional diseases as heart problems, kidney ailments, and cancer. In that climate, Bigelow became intrigued by the phenomenon of hibernation, a strategy necessarily adopted by many animals lower down the food chain in his frigid part of the world. During the lengthy winters there certain animals survived for months without food by significantly slowing down their hearts.

Taking this page from nature, Bigelow cooled down the body temperature of laboratory dogs, cut directly into their hearts, and operated for more than four minutes—and they survived. Why not apply this technique of cooling, or hypothermia, to the human patient?

At the University of Minnesota in Minneapolis, Clarence Walton Lillehei and John Lewis enthusiastically agreed that Bigelow was on the right track. For about a year they performed "open-heart operations" on lab animals, using hypothermia.

Finally, on September 2, 1952, they were prepared to try to save the life of five-year-old Jacquie Weeks, who was rapidly dying from a congenital defect, a small hole that allowed blood to leak between the heart's chambers. Only such a desperate case could rationally be attempted, since the risks of using hypothermia were not slight. If the heart was cooled down too much, it might stop dead, never to be restarted. Lillehei later described the tense beginnings of this historic event: "We wrapped her body in a special rubberized blanket containing tubes through which a cold alcohol solution flowed. By cooling Jacquie to a temperature of 81 degrees F, [we could help her] survive without a pumping heart for about 10 minutes."

As the child's heart slowed down, beating in slow motion, the surgical team clamped shut the veins flowing in, draining the heart of blood. Quickly they repaired the hole that threatened her life. The whole nerve-racking procedure took five and a half

Used from about 1960 through about 1963 for open-heart surgery at New York Hospital–Cornell Medical Center, the Hays-Cross oxygenator was one of several landmark machines that temporarily took over heart and lung functions, thus extending the safety zone for surgical interventions. Shown here are the control unit and the coils of the mechanism. First invented by John Gibbon in 1951, heart-lung machines performed two functions: pumping the blood in order to keep it circulating through the body's great vessels, and oxygenating the blood with an artificial "lung."

minutes, an unprecedented length of time for a patient to survive without blood coursing regularly through the brain.

When Jacquie's heart started beating again, Lillehei relaxed and said to Lewis: "This is it, John. We're into the heart to stay." It was a defining moment in modern surgery.

To bring the girl's temperature back to normal, she was immersed in warm water. There was no brain damage, the heart worked normally, and she went on to live the life she chose, eventually having two children of her own.

"200% Mortality Rate"

Still, the outer limit of the hypothermia method was ten minutes—not nearly enough time to repair heart defects more extensive or complex than Jacquie's.

In a desperate experiment that no hospital ethics committee would allow to take place today, Lillehei began trying to extend a young child's open-heart operation past ten minutes by attaching his blood vessels to his parent's. The idea was to have the parental heart and lungs pump blood for the child. On one occasion, as the operating team watched in horror, an air bubble sped through the connecting clear plastic tubing into the parent's veins, causing a fatal heart attack. Nor did the child survive.

Lillehei decades later said, "I will go down in history as the only surgeon to invent an operation with a 200% mortality rate." In fact he balanced a child's certain death from heart malfunction against the risks of this procedure another forty-six times; the survival rate of the desperately ill children was 63 percent, with no further deaths of participating parents.

Meanwhile all kinds of alternative ideas were being tried out by creative surgeons and medical technicians. The so-called pump oxygenator machine, about the size of a concert grand piano, was the first serious attempt to build a device that would take over the function of the heart during an operation. It was imperfect, but it inspired more sophisticated successors.

One of them became a national prime-time live television star in 1958. The bubble oxygenator, which was much less cumbersome than the pump oxygenator, was used by California physicians in a televised open-heart operation on a little boy named Tommy. From 7:00 to 11:00 P.M., as the surgical team worked its magic to save his life, the real-life drama drew higher ratings than any other program, including *Cheyenne*, the day's most popular regularly scheduled show. Not so incidentally, this effort, along with coverage in the print media of developments in open-heart surgical interventions, caused the average American's opinion of modern medicine to soar sky-high. Magic bullets inspired the popular press to place physicians on pedestals. Open-heart surgery became reason to affix wings to their backs. The public's expectations began to outpace reality or reason, in part because of campaigns orchestrated by organizations of medical professionals.

Two Pathways

In addition to heart-lung devices, there were many machines in operating rooms, beside hospital beds, and in doctors' offices at mid-century, from X-ray to electrocardiogram machines, but one thick, sullen, groaning machine struck terror in most hearts. The iron lung, a large cylinder in which paralyzed victims could lie until the end of their lives, did the breathing when the chest muscles that propelled the heart no longer worked.

To anyone in the 1950s the iron lung was a grisly symbol of poliomyelitis, a crippling and incurable illness caused by a virus. Afflicted children wearing clanking and uncomfortable braces, adults who like Franklin Delano Roosevelt were crippled for life,

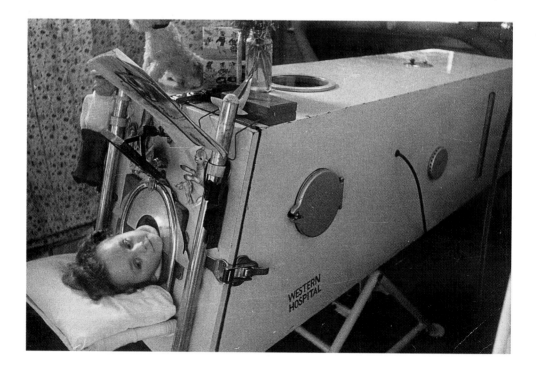

horizontal patients with little but their heads visible at the end of their sustaining iron lungs: All seemed to be irrevocably disabled.

Unlike bacterial infections, viral infections are not susceptible to cure even in our own day. The viruses that cause polio and many other diseases were discovered during the first decade of the century. Much smaller than bacteria and invisible to humankind until the development of the electron microscope in 1932, they can reproduce themselves only by invading other living cells and seizing their nucleic mechanisms.

The virus, like other organisms that invade the body, is covered with antigens, or characteristic molecules that have an identifying shape. To block these antigens, special cells in the immune system speedily reshape themselves and hold the enemy molecules in place. These antibody cells keep watch until the body's killer cells race up to wipe out the foes. This is an excellent plan, but it does not always work. Some viruses can mount blitzkriegs, multiplying lickety-split and overpowering body cells. Even so, the immune system usually prevails. Ever afterward the forewarned body is forearmed with hosts of antibodies on the prowl for that particular assailant.

One avenue for preventing a viral illness therefore is to stimulate the immune system to produce the precise type of antibody necessary to repel the causative virus. A vaccine that gives a mild dose of the disease will create enough antibodies to deal with a deadlier invasion.

The concept of vaccines goes back to the eighteenth century, when the English surgeon and naturalist Edward Jenner noticed that milkmaids who had cowpox blisters on their hands never contracted smallpox, a much more acute and often fatal viral disease. Taking something of a risk in 1796, Jenner scraped infected matter fresh from the lesions of a cowpox-infected milkmaid, made a vaccine, and injected James Phipps, a healthy eight-year-old. The boy suffered a mild case of cowpox but bounced back to health. Then Jenner boldly injected Phipps with smallpox, and there was no reaction. Jenner correctly reasoned that there was an eloquent connection here; indeed, as we now know, the two viruses are similar enough that antibodies formed in

First developed in 1927 by Philip Drinker and Louis Shaw, the mechanical respirator known popularly as an iron lung kept severely paralyzed polio victims alive. Except for the head, the victim's entire body was encased in a rigid metal tank that alternated positive and negative pressure to provide artificial ventilation. In the polio epidemic in the United States from 1943 to 1956, some twenty-two thousand victims died out of the total four hundred thousand infected.

All viruses use similar mechanisms to attack human cells. The adenovirus shown in this false-color scanning electron micrograph, magnified fifty thousand times, causes an infection of the upper respiratory tract with symptoms like those of the common cold. Projecting from the virus's protein coat are spikes with which the virus recognizes a potential host cell. The virus has just enough DNA in its core to commandeer the cell's genetic machinery in order to replicate itself.

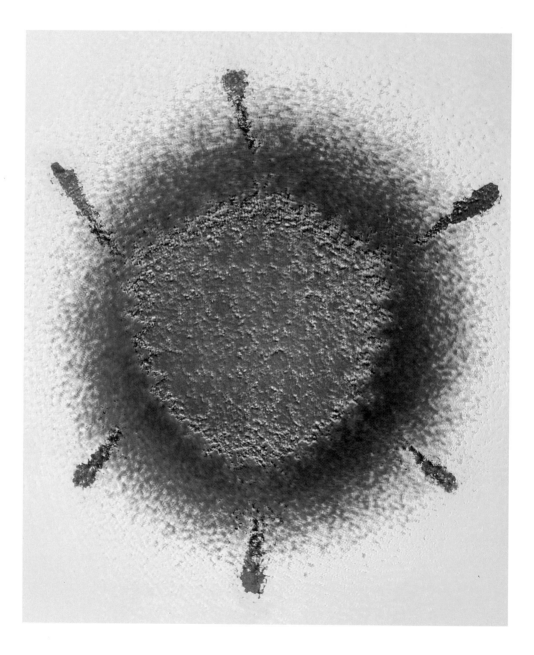

response to a cowpox infection are well able to hold the line against smallpox. The result was so simple, so compelling, that after Jenner published his findings in 1798, Thomas Jefferson exuberantly proposed that everyone be vaccinated with cowpox. In a democracy the idea was intelligently debated, even as the death rate from smallpox fell dramatically, but no serious action was taken. In France the emperor Napoleon decreed that every soldier in the nation's army be vaccinated, and that was that. Except for mishaps when physicians did not precisely follow Jenner's directions, the disease was on the run.

Many experiments with vaccines in the eighteenth and nineteenth centuries did not turn out well, usually because researchers did not have the skill or equipment to isolate the required antibodies from other elements in the blood serum. Even as late as the 1920s results were unpredictable. Working at the Pasteur Institute, for example, the French biologists Albert Calmette and Camille Guérin tried to develop a vaccine against tuberculosis, then the most prevalent and deadly infection in the world. It is caused by a bacterium, but they used the same method that would work with viruses. Using serum from cattle afflicted with bovine tuberculosis, they came up with a drug called Bacillus Calmette-Guérin (BCG), but its successes were few.

This record did not bode well for an effective, reliable, nontoxic vaccine against polio, a disease that became epidemic only in the twentieth century. One memorable outbreak startled and frightened Americans in the summer of 1916, hitting with particular force in highly populated New York City. Then and later there seemed to be some connection with the hot months—even today no one knows why—and with crowded public places like swimming pools and movie theaters.

Distraught parents forced their children to stay inside hot, stuffy houses during the months of greatest danger. There were, predictably, speculations that poverty, dirt, and foreign influences underlay the epidemic and that children were primarily at risk. By the 1920s, however, as the case of adult, native-born, wealthy, and presumably well-bathed Franklin Roosevelt proved resoundingly, the speculations produced unintended consequences. Like other children of privilege, he was protected from exposure early in life, when infection might have been milder and given him immunity as an adult.

In fact, cleanliness was a serious risk factor, rendering the emerging middle class especially susceptible. The explanation recalls the cowpox/smallpox model. A form of polio about as prevalent and dangerous as the common cold was often encouraged by unclean conditions. It passed quickly, leaving antibodies that could fight off the more virulent, crippling form of the disease. In sum, many victims lacked antibodies because they were geographically and socially isolated from urban areas.

By World War II a little more was known about the behavior of the polio virus. It was found to take up original residence in the intestines, mutating and becoming stronger there before migrating to attack the spinal cord. While biding its time, it might be transmitted out of the body in feces. Thus sewage could infect rivers and oceans with the polio virus, and the virus could be passed along by infected people with unclean hands or by mothers changing diapers.

Despite this new understanding, the epidemic appeared again and again in terrifying waves, most memorably in 1952 and 1953. Though still popularly known as infantile paralysis, it could be much more devastating to adult victims. In some cases the virus might travel to the nervous system and affect only a single routing of nerves, most

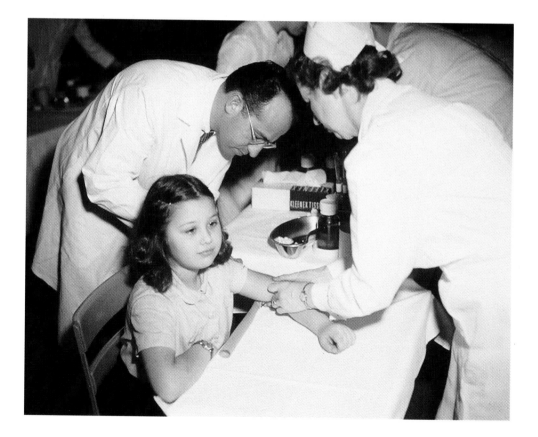

often paralyzing a leg. In others, as in the cases of those who lived within iron lungs, the result was virtually total paralysis.

After the war the nation could marshal greater resources of money, talent, and technology in an aggressive campaign against this cruel virus. Basic research was funded by the National Foundation for Infantile Paralysis, whose extraordinarily effective fund-raising effort was well known as the March of Dimes.

In 1948 John Enders sped up the race to discover a polio vaccine by growing the virus in culture. Inadvertently he also parboiled the intense rivalry between two ambitious, self-assured scientists who both—apparently to their mutual displeasure—became recognized as giants in the field of viral medicine. The two men set out on two clearly distinct pathways of research: the "killed" virus and the "live" virus.

Jonas Edward Salk, who headed the Virus Research Laboratory at the School of Medicine at the University of Pittsburgh, favored the former. He believed that the safest approach was to engineer a strain of polio virus that would be recognizable by its surface of characteristic antigens and thus stimulate the body's immune system to produce antibodies. But he wanted to destroy the virus's genetic material, producing the so-called killed virus, so there was no possibility that it could replicate.

Albert Bruce Sabin disagreed. Mimicking to some extent the natural immunization stimulated by the mild form of polio encouraged by unclean conditions, he preferred to base his vaccine upon a virus that was sufficiently live to spark a response from the immune system but too weak to cause the full-blown disease.

A major obstacle for Salk was that the polio epidemic was caused by three separate strains of poliovirus. He used formaldehyde to kill samples of all three types, then produced a vaccine from the combination of killed viruses and injected it into laboratory monkeys. The first tests were promising: Antibodies formed in the blood of the animals, and none contracted the disease.

Unlike experimenters of earlier decades, the Salk team moved ahead cautiously.

Dr. Jonas Salk, an American virologist, supervises an early inoculation with his polio vaccine at the Sunnyside School in Pittsburgh in 1955. He had developed his so-called killed virus vaccine three years before, but it was not ready for use in mass inoculations until 1954. In 1960 Albert Sabin developed a live-virus vaccine that became even more widely used. Vaccination caused the disease to disappear entirely from the Americas by 1994. By the year 2000, some say, poliomyelitis will exist nowhere on earth.

The vaccine was tried first on afflicted children who had developed an immunity to the virus, then on subjects, including some members of Salk's family, who had never been exposed. Again and again the vaccine proved successful. In 1954 almost two million children were injected with either the Salk vaccine or a placebo. The vaccine was effective in more than four out of five cases.

The Salk vaccine was introduced into the United States in the largest vaccination campaign in history. Unfortunately the rush to manufacture the first effective poliomyelitis vaccine had a terrible side effect: One improperly produced batch caused several hundred cases of the disease.

Meanwhile Sabin's version was not ready in time to compete. In a move unparalleled during the Cold War era, this American physician took his polio vaccine to the Soviet Union and distributed it to eager millions. Dipped in sugar cubes and taken by mouth, it proved to be much more user-friendly than Salk's painful injections. It was cheaper too. But it became the vaccine of choice in the United States by the early 1960s for medical reasons: Experts came to believe that the live virus fooled the human immune system into manufacturing stronger, longer-lasting antibodies than did the killed virus.

Today's parents have a choice. The Sabin vaccine produces some dozen cases of polio each year, though only when a child's immune system has somehow been compromised beforehand. Also, caregivers and parents contract the disease at about the same rate. Generally, the Salk version is considered effective enough, so long as the risk of infection from polio seems very small. But in areas where the disease has reappeared in significant numbers, physicians will probably recommend the Sabin live virus vaccine.

Still, eradication of the disease would be the happiest resolution of the polio story. Even today thousands of adults who recovered from polio before the invention of either vaccine live in fear. Postpolio syndrome, a debilitating disease, can often strike decades after a childhood bout of polio and is probably its long-delayed, unpredictable legacy.

The Two-Headed Dog

While the world marveled at the achievements of bacteriologists and virologists at midcentury—in short, as millions of people realized that they and their loved ones were alive and whole only because of what popular culture saw as miracle drugs—surgical researchers had been no less active or creative.

From the earliest years of the century they had known that, given the proper techniques, it might be possible to transplant human organs. Drs. Alexis Carrel and Charles Guthrie experimented with grafting the organs of one animal onto another. In 1905 they created a sensation with their two-headed dog. The subjects of their experiments inevitably died, but the two men were able to perfect surgical techniques that others could emulate.

By the 1950s Dr. William Kolff had designed an artificial kidney, a dialysis ma-

Dr. William Kolff wears his groundbreaking device, the first "artificial kidney" or renal dialysis machine, which he developed during World War II while working in his native Holland at the Kampen Municipal Hospital. Fashioned from lightweight aluminum and wood, it reproduced the function of healthy kidneys by using several cellophane membranes to filter impurities out of the blood. All renal dialysis machines in use today are modeled upon this invention, which helps prolong the lives of patients whose kidneys have failed completely or gives damaged kidneys a rest while they heal.

chine that filtered waste products out of the blood. It completely cleansed a patient's blood within fourteen hours, but the effect was temporary. For patients whose kidneys had permanently failed, it provided tedious, painful maintenance. But Kolff's invention laid the foundation for kidney transplantation, just as early heart-lung machines had made open-heart surgery possible.

Francis Moore took the first cautious steps in making organ transplantation almost a pedestrian procedure today. Chief of surgery at Peter Bent Brigham (now Brigham and Women's), a small hospital in Boston, he began with three patients dying from kidney disease or malfunction and three healthy kidneys harvested from recently diseased individuals.

The chief obstacle, Moore thought, would be the likelihood that the patients' bodies would reject the strange new kidneys as foreign invaders and attack them with legions of defensive white blood cells. Among other stratagems, Moore therefore decided to try wrapping one of the kidneys in a plastic bag in order to protect its outer surface from assault.

Only one patient survived, a twenty-six-year-old South American physician afflicted with chronic glomerulonephritis, a disease affecting the network of capillaries in the end capsules of each kidney tubule. Nineteen days after the transplant his new kidney was fully operational, pouring out urine. Although he lived only another six months, his recovery had proved that the operation could be successful.

Truly Identical

In most cases kidneys were being rejected, plastic bag or no. Then, in 1954, Moore and his surgical team were visited by a desperately ill patient who, they immediately recognized, was unusually likely to benefit.

Richard Herrick, twenty-two, was dying of Bright's disease, but a healthy kidney could save him. Unlike any previous patient, he had an identical twin brother. The surgeons reasoned that Richard's body would not consider his brother's kidney a hostile intruder because the match was identical.

Eight years later Richard finally succumbed to Bright's disease, but his life had been substantially extended by the kidney transplant. Other identical twins could expect the same fine results, but what about the rest of the world's kidney patients? It was now clear that the twenty-six-year-old South American had survived only because of a fluke match.

The Brigham researchers cast wildly about for potential answers. Because radiation from the atomic bomb blasts at Hiroshima and Nagasaki had wiped out the immune systems of some survivors, they tried that path. That is, they gave each transplantation patient a heavy dose of full-body radiation, thus destroying the body's ability to reject the alien organ. Unfortunately the irradiated body was rendered totally vulnerable to infection. In one case, since sterile isolation rooms had not yet been established as part of standard care, a woman had to be kept in the operating room for twenty-eight days. She escaped any life-threatening infection but bled to death. The radiation treatment had destroyed the coagulating agents in her bloodstream.

The Mirage of Immunosuppression

Then the picture brightened considerably. In 1959 George Hitchings and Gertrude Elion, researchers at the Burroughs Wellcome laboratory, synthesized azathioprine, the first truly promising antirejection drug. Harvard University's Joseph Murray and Roy Calne seized upon this new immunosuppressant compound for their experiments with

Participants in the first successful kidney transplantation, the identical Herrick twins were so closely matched genetically that the transplant recipient's body did not reject the donated kidney of the other. Performed in Boston in 1954 by a surgical team that included Joseph Murray, John P. Merrill, J. Hartwell Harrison, and Warren Guild, the operation is performed frequently today, but a transplant can still be unworkable when the donor organ does not provide an appropriate match with the patient's tissues. Researchers are now hoping to develop an artificial kidney that can be surgically inserted into the body.

dogs and were cheerfully astounded. Azathioprine did not apparently cause much collateral damage to the body. Better still, the survival rates of dogs with transplanted kidneys began averaging longer than a year.

But the problems of deceiving the more complex human organism were not so easily resolved. The first two patients given azathioprine were poisoned to death. Doctors had to experiment to determine the safe level of dosage of this powerful new agent.

For the third attempt Harvard's surgeons chose a twenty-four-year-old near death from Bright's disease. Team members mounted a round-the-clock deathwatch at the hospital, taking turns sleeping there in order to be on hand when accident or illness carried off a patient with a healthy kidney to donate.

In April 1962 a kidney at last became available, and the operation took place, using about a quarter of the dose of immunosuppressant given in the two failed transplantations. For the moment the signs were spectacularly good. Four months later, however, the patient experienced a rejection crisis, which was conquered with cortisone. Next, he had to be given antibiotics to fight off pneumonia. These difficulties were only the beginning of a very rocky road. Eighteen months later the patient developed acute appendicitis; meanwhile there were signs that his body was going into chronic rejection of the new kidney. The surgeons were able to rustle up a second kidney and successfully transplant it. Six months afterward the patient died from a liver infection.

This exhausting series of tangled events left Harvard's researchers puzzled. Had the liver been damaged by the toxicity of the immunosuppressant? Or had the patient been infected with hepatitis from a blood transfusion?

In the end the mystery became academic. Once Moore and his team published the results of their tantalizing near success, surgeons around the world realized that fully triumphant transplants were within their grasp. The barrier of immune system rejection had been breached for the first time, if only briefly, by the use of azathioprine. As with previous breakthroughs in the use of drugs, experience and improved methodologies would consolidate the advance. It was still better, however, to have an organ from a live related donor, a parent or a sibling.

It soon became clear that the kidney was only the first vital organ that could be replaced in order to extend human life. In 1964 the first liver transplant was attempted by Drs. Tom Starzl and Francis Moore. Dr. James Hardy become the first surgeon to transplant a lung. It was only natural, as heart disease became a major killer, that imaginative surgeons began wondering just how insane it would be to try transplanting the body's complex "forbidden organ."

"Cut Well ..."

In South Africa the flamboyant cardiologist Christiaan Barnard extracted a beating heart from a brain-dead woman on life-sustaining equipment, then inserted it in the dying body of Louis Washkansky, a fifty-four-year-old Johannesburg grocer. This so-called Miracle of Capetown grabbed headlines worldwide and sparked a virtual frenzy of heart transplantation, even though the miracle ended eighteen days later for Washkansky.

His death was in no way regarded as reason for despair. Barnard's colleague Adrian Kantrowitz tried a transplant on a fatally ill eighteen-day-old infant. He failed. Barnard gave a new heart to fifty-eight-year-old Philip Blaiberg, who survived for nearly two years. East across the Pacific, Stanford University's Norman Shumway performed the world's fourth known heart transplant, but the patient did not last long. In Houston, Texas, Denton Cooley was even more daring, transplanting a heart and lung into a two-month-old. The child survived only briefly. But Cooley was persistent. Eventually, after performing twenty-two heart transplants, he could explain his secret in no-nonsense terms to an arsenal of broadcast mikes and cameras: "Cut well, tie well, and get well."

With so many transplantations, failed or successful, a problematic consumer shortage occurred. Understandably few members of the public signed up to donate their hearts upon their deaths. The whole process was not widely understood. Even among professionals and in law there was not yet general agreement on how to proceed or even on the exact definition of death. In some states, as in South Africa, a heart could be harvested legally from a brain-dead patient. In others the law had determined that death did not occur until the cessation of the heartbeat. This kind of stricture made it difficult to gather still-healthy hearts in a timely manner.

In frustration James Hardy resorted to using a chimpanzee heart for one of his patients, but the recipient died within hours.

Then the leaders in the field began to look into the potential of mechanical devices. In 1969 Cooley could not find a donor heart for his rapidly sinking patient Haskel Karp. In some desperation he decided to make use temporarily of a new experimental machine, perhaps unfortunately known as a mechanical heart. This is the kind of name

that can raise unrealistic expectations and muddy public understanding of complex issues and therapies. That is exactly what happened.

After Karp had been attached to the machine, his wife went on national television to appeal for a donor heart. A suitable organ turned up, Cooley performed the transplantation, and Karp died two days later of pneumonia. In the scandal that resulted, Cooley had to resign his hospital post. Mrs. Karp brought suit against him for experimenting with her husband, but she was unsuccessful.

In the aftermath there was more outrage as the public learned the truth about

Left: After the successful transplant of a donor heart, Fredi Everman holds his own failed organ in a jar of preservative. Low survival rates due to rejection cooled the enthusiasm for heart transplants in the late 1960s and early 1970s. These problems were essentially solved a decade later, however, and the operation is considered unremarkable today. Below: Developed by William Jarvik, this mechanical device, once implanted into the patient's heart cavity, uses compressed air to pump blood, theoretically reproducing the essential function of the human heart. Its sensitive mechanism adjusts automatically to ordinary variations in the flow of blood. Despite partial success in more than a dozen operations, the device is no longer in use because of circulatory problems that have not been solved.

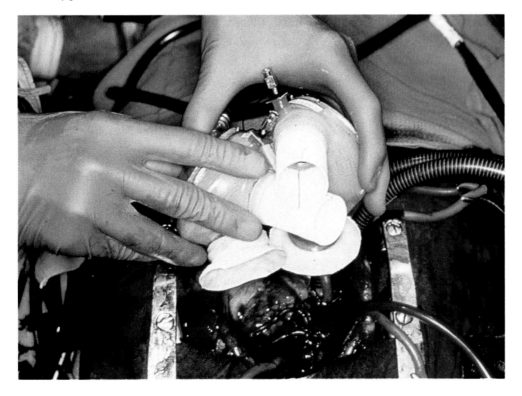

heart transplants, which had been oversold. As few institutions had rushed to reveal, rejection of foreign hearts as well as most other donor organs was still a huge, seemingly intractable obstacle. The average survival rate was only about twenty-nine days. Laypersons became seriously disillusioned with the whole effort, the number of operations dropped drastically, and Cooley entirely abandoned the procedure.

Meanwhile medical science was concentrating on another major rampaging killer.

Clusters of Lung Cancer

Cancer has been familiar to humankind at least since the earliest-known writing. No cure has ever been found. Treatment throughout our own century, which has engendered so many wonders of medical discovery, has been based on the most basic of principles: Cut and burn. In other words, the tumor has been either surgically removed or burned out with radiation and/or chemicals.

Resistant to the finest scientific minds, the most concentrated team efforts, and oceans of private and government funding, cancer has perhaps provided one slight benefit: It has been a reality check on anyone tempted to believe blindly in the infallibility of modern medicine.

But if no one has a cure, there have been many studies trying to establish links or causes. When U.S. Surgeon General Luther Terry announced his famous warning against tobacco in 1964, he was relying on a wealth of excellent research, but the most conclusive had been conducted by Austin Hill, a British epidemiologist. Along with his partner, Richard Doll, he managed the first large-scale studies to establish a link between cigarette smoking and cancer. Taking note of clusters of lung cancer in certain sectors of the population who had come into greatly increased contact with automobile traffic, they speculated that road tar and exhaust fumes might be responsible.

Doll's methodology was to retrieve extensive personal histories of the cancer victims in highly detailed questionnaires. To his astonishment, an unforeseen pattern emerged with startling clarity: A huge percentage of the patients were longtime smokers. Spurred by this finding, the team began critical prospective studies—that is, they chose a group of physicians who smoked and followed the progress of their health into the future. Yet the link was already clear. These studies became the accepted survey and statistical model for determining connections between environmental factors and disease.

Meanwhile many other researchers began looking inside the body, which had

The malignant characteristics of cancer cells in the human body are partially visible in this cross-section of a single cancerous cell, photographed eighty thousand times actual size. The dark blue material is genetic DNA. Rather than forming a complete nucleus, it is randomly distributed throughout the cell, which has a characteristically irregular surface. A daughter cell retaining these defects has been formed at the upper right.

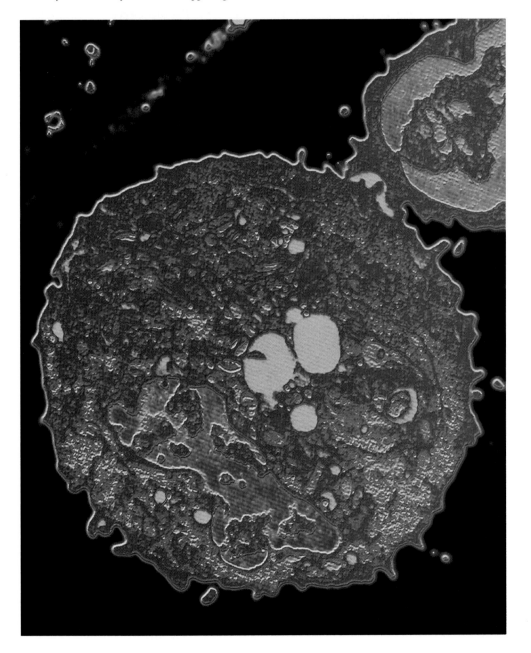

often before been found to be treacherous. Michael Bishop and his colleague Howard Varmus, for example, thought there might be some innate factor that triggered the unregulated growth of cancer cells. In fact it was known that these cells were somehow receiving mixed-up signals from genetic material.

Then, in addition to this problem and the environmental causes, there arose a third problem. A researcher named F. Peyton Rous had been able to cause cancer in chickens by contaminating them with a virus that became known as the Rous sarcoma virus. The question now was: How did these three triggering factors interact?

Doll recently recalled the situation:

> *Smoking was obvious. So, too, were materials and chemicals like asbestos. And we knew that certain cultures had out-of-average rates for particular cancers. The Japanese, for example, had an abnormally high rate of stomach cancer . . . so what we did or did not take into our bodies clearly affected our chances for getting the disease. And hepatitis B virus was clearly linked to incidences of liver cancer, and papilloma virus to pelvic cancer.*

In the mid-1970s Varmus and Bishop isolated the genetic material that seemed to be linked to the cancers in Rous's infected fowl. Of the virus's four genes, they discovered, one could insinuate itself into the DNA of a chicken cell and cause it to begin dividing. The big discovery was that the cancer-causing gene also existed in other creatures, including humans. This SARC gene was slightly mutated; whenever the mutated form was found, it was shown to be acting as a trigger for cancer. Varmus and Bishop called this mutated gene an "oncogene." Since their discovery, about a hundred oncogenes have been found. Such environmental factors as cigarette tar and toxic chemicals can cause a healthy gene to mutate into an oncogene, a transformation that is the first step in a process that can cause cancer. Another class of genes, known as tumor suppressor genes, can also initiate that process.

Because both types of genes can be inherited, cancer is sometimes a genetic disease. Those who inherit oncogenes or tumor suppressor genes are given a jump start on succumbing to cancer, typically showing signs of the illness in childhood or by their thirties. Evidently about 5 percent of cancers are passed on in this way.

The other 95 percent occur either because of damage that produces oncogenes or tumor suppressor genes or because, in the ongoing growth of the body, mistakes simply happen during replication. In either case, one mutation event is generally not sufficient to produce cancer. Over several decades, several separate mutations are required.

In 1971 U.S. President Richard Nixon optimistically declared "war" on cancer, then killing about one thousand Americans every day. Since then a few cancers—but not the deadliest—have been defeated, but it is estimated that cancers will be the leading cause of death in the United States at the century's close.

In a Norwegian Soil Sample

One important exception is childhood leukemia. In 1950 diagnosis of this disease meant that a child would die within three months. Ten years later researcher Sidney Farber found a way of inducing remission with drugs that contained folic acid blockers. But this proved to be merely a stopgap measure. The cancer always returned and inevitably killed.

Few enigmas in medicine inspire such frantic research as a fatal disease that strikes young children overnight. At the National Institutes of Health researchers tried several approaches, from controlling infection to analyzing the effectiveness of chemotherapies to dealing with the hemorrhaging associated with leukemia. In addition, as very sick children lay dying in the NIH hospital facility, their bone marrow was tested daily to see what was occurring.

Sometimes a drug seemed promising; then the recovering patient relapsed. Despite fears of the consequences and questions about the ethics involved, the researchers sometimes tried different combinations and various dosage levels of experimental drugs. Nothing worked well.

To be lastingly effective, any treatment had to eradicate every last leukemia cell. The average victim was host to ten trillion of them, an invading force weighing about a kilogram. One researcher working with leukemia-ridden lab mice found that it took no less chemotherapy firepower to kill the final surviving milligram of cells than to destroy the rest of the kilogram. For that reason, treatment teams calculated that drug treatment should be continued for at least 164 days. This regimen produced remissions but no cures.

By the 1970s the NIH experiments were testing the potential of fighting leukemia by completely replacing the children's bone marrow. First, the affected bone marrow was destroyed by radiation and chemotherapy. Next, bone marrow from a healthy donor was injected into the patient's vein. Why did the victim's body accept this benign intruder and not call up the rejection brigade?

Because a chance discovery had fortuitously yielded what labs could not produce: In a sample of Norwegian soil a previously unknown substance had been found that profoundly suppresses the body's immune system. Cyclosporine was soon used for other transplant procedures as well as the bone marrow operation. Survival in liver transplants, which had been virtually abandoned, leaped from 30 percent to 70 percent.

Who Decides?

By the end of the century this natural immunosuppresant ensured that rejection in a heart transplant was no longer inevitably followed by death. Nine out of ten times such rejection responds to medicines taken at home. In American hospitals the survival rate after a heart transplant has become 80 percent for the first year after the operation, 60 percent for ten years.

But each increment of improvement aggravates the demand-supply problem. In the United States alone forty-five thousand patients could benefit from heart trans-

The amazing strides made in organ transplantation since mid-century have not resolved all of the issues. The growing demand for harvested organs far exceeds the supply. One possible solution is to harvest healthy body parts from transgenic pigs—that is, pigs whose genetic material contains human genes. The subject here is involved in research on animal-to-human transplantation, known as xenotransplantation, on a farm operated by Baxter Healthcare Corporation. Genetically altered pig livers are currently being tested as an external support system for patients awaiting a human donor organ.

plants every year, but only twenty-two hundred of the procedures, which can reach five hundred thousand dollars in medical and hospital costs, are performed annually. A heart patient is now more likely to die while on the waiting list than in the first twenty-four months after receiving a transplant. Currently another avenue is being explored: Attempts are being made to breed pigs whose organs will be serviceable in the human body because their genetic material includes human regulatory proteins that protect the animal organ from hyperacute rejection.

As with many radical advances in medical treatment, new issues of safety and ethics are raised with each new chapter of the transplantation story. Safety precautions today, zealously scrutinized by the federal government, are much stricter even than in the recent past. No surgeon could take the risks of using a mother's circulatory system as a fail-safe system during her child's open-heart surgery or of injecting staff nurses with an untested substance in the interest of pure science. Applications have to be made, and oversight committees convinced. To take one example, surgeons have been given the green light to use pig livers as bridge organs until a human donor organ is available. But this particular organ has been deemed acceptable only for this specific use.

Ethical questions are even more difficult—in part because they potentially affect a multitude of belief systems in a diverse society, in part because idealism necessarily comes into conflict with brutal realities, including the rise of corporate medicine. Priorities in medical care change radically, of course, if sustaining a healthy profit margin becomes as important to hospital administrators as restoring or maintaining the health of a patient.

A transplant operation is enormously expensive, maintenance of the patient on the immunosuppressant cyclosporine costs up to ten thousand dollars a year, and the extension of life may total only a few months or years. Because the body's immune cells are formed in the bone marrow, transplants of bone marrow from a healthy donor to an afflicted patient are frequently successful. Occasionally, however, the procedure backfires, producing graft-versus-host disease. Cells from the transplanted marrow attack the patient's body after an expensive investment of limited resources.

Is such a huge outlay of money, hospital facilities, and professional help practical in the cost-effective sense? Especially if the society as a whole must bear the cost of these treatments? Or if large corporations feel no shame in believing that amassing optimum profits is more important than paying for their employees to take advantage of such treatments? And if the patient's own choices—to smoke, to drink heavily—have certainly contributed to the deterioration of a vital organ? What about the need to prioritize? Does the most readily available donor organ go to the patient closest to death or to the one most likely to survive a long time? Should the patient with bags of money, whether inherited or earned, be given precedence over the patient with no medical insurance or other significant financial resources to contribute? Is the parent of young children more "deserving," in this special social sense, than the flighty young socialite or the shy, retiring loner? These are highly charged questions, but they will have to be boldly addressed, barring some unexpected market infusion of animal or mechanical organs in large numbers.

Such issues will become even more pressing as transplantation becomes ever more inventive. Some experts predict that within decades it will be possible to replace limbs.

If that sounds implausible, consider what our great-grandparents would have thought if promised today's medical landscape. Antibacterial drugs and viral vaccines have saved tens of millions of lives. Smallpox, an ancient and aggressive killer, has been wiped off the earth. In the operating room, coronary artery bypasses save tens of thousands each year from predictably fatal heart attacks. New human life begins in the test tube.

Then in San Francisco, as in the epidemic that came to light with the death of the plague victim Gin, an even more terrifying and refractory contagion was discovered in the last quarter of the century.

The "Perfect" Virus

It began with a mystery. Why were a few young men dying from a rare form of pneumonia that had previously been seen only in elderly patients with immune deficiencies? Only after an initial period of denial by some experts and government officials did it become apparent that an invincible new epidemic killer was on the loose. At first the viral infection that became infamous as autoimmune deficiency syndrome, or AIDS, was visible only in the city's homosexual community.

Eventually researchers discovered exactly how the virus works. Like all other viruses, it enters through the cell wall, takes over the nucleus, then destroys the cell for its own nourishment. But because this intruder wipes out immune cells, its human host becomes susceptible to any opportunistic infection.

Rising from a host T-lymphocyte white blood cell, HIV viruses involved in AIDS have acquired a membrane, falsely colored green in this electron micrograph, from their host's membrane. The departing viruses were formed within the T-cell, which is part of the body's immune system, after it was injected with viral RNA genetic material by a parent virus. The damage to the T-cell membrane may kill it, thus compromising natural immunity.

"AIDS doesn't kill you," explains Frederick Murphy, one of the world's leading authorities on emergent disease. "It beats you down to the point where other microbes finish the job."

With grim humor, researchers dub AIDS the "perfect" virus, for its strategies are devious. The immune cells it attacks are the only defense specifically designed to destroy it. It endures within the body for years, giving its host opportunity to infect countless other victims. And it is primarily transmitted through sexual contact or intravenous drug use, two activities that are not easy for their practitioners to stop.

Yet altered behavior, such as using condoms, is the only major protection possible today. Unfortunately, as Murphy puts it, "Young people think they are going to live forever . . . and drug users don't think very clearly at all."

In a limited way progress has been made as private and government resources have become partners in trying to explain and defeat this terrifying virus. On a regimen known as three-drug therapy, several patients have tested negative for the virus for as long as three years. But researchers cautiously assume that this is not proof of cure; more probably the virus is being held in check by the complex treatment, which costs about twelve thousand dollars a year, and the disease has become a chronic condition for the patient. To date, treatment seems to be most effective if started as soon as possible after infection is discovered.

Yet again money has become a major player on the treatment team. In the early days of the AIDS epidemic most patients in the United States were college-educated, white, middle-class, homosexual men who usually had insurance or other financial resources to pay for care. That profile has altered greatly. Today's victim is likely to be poorer, younger, darker. AIDS is now discovered more frequently among IV drug users, and it is increasingly apparent among heterosexuals, especially in minority communities.

Two Kinds of Health Care

Such shifts spotlight a long-standing reality of American health care that is only getting worse: The poor, whatever their ethnic backgrounds, receive less effective health care

Vaccination of children in Ethiopia is part of a World Health Organization attempt to prevent common diseases in less developed countries by providing basic health services. The stated goal of the World Health Organization is "complete physical, mental, and social well-being, and not merely the absence of disease."

than do more comfortable members of society, while African Americans suffer the most severely from declining medical resources in their communities. White infants have a 70 percent higher likelihood of reaching their fourth birthdays than do black babies. A male adult in Harlem has the same statistical chance of living past forty as a male adult in Bangladesh, one of the world's poorest nations. Across the board, whites live longer than African Americans, are less susceptible to heart attacks and cancer, and are less likely to die from infectious disease.

For the poor there is little practical difference between city and country in these matters. Some public hospitals in large, teeming urban areas are barely surviving as functioning institutions. In some rural areas a patient may have to travel hundreds of miles to come within range of treatment, then be refused because of lack of cash, Medicare, or Medicaid.

It follows that preventive care—the Pap smear, mammogram, cholesterol screen—is mostly out of reach. Millions of Americans of all ages cannot afford to practice the best kind of medicine, prevention of illness. Instead they wait until a crisis,

when they are forced to resort to the expensive care of an emergency room, and that puts strains on the entire health care system.

"Health care for the poor is in crisis," warns medical historian and physician Vanessa Northington Gamble, "and it's only going to get worse. There's something very wrong with our system, our priorities, when we have many communities in the very shadow of well-equipped, high-tech hospitals dying of things they shouldn't be dying of."

In Third World countries, so-called small medicine, such as the widespread use of vaccines and information about preventative medicine, is extraordinarily successful in dealing with infectious disease. The big-ticket medicine offered to the rich or government-subsidized in the industrialized world is not necessary to raise the prevailing quality of life to surprisingly high levels. It is possible that the small-medicine model could be very effective in many American communities as well.

But professionals agree that there are many more challenges on the unpredictable journey ahead of us, especially in the study and treatment of chronic disease. For everyone it has become increasingly important to learn more about taking charge of one's own health future, by eating less or better, by exercising more, and by cutting back or ending all provably self-destructive activities.

Will the future discoveries of genetic scientists finally unlock the mysteries of cancer and other fatal or debilitating diseases?

In 1994 the professional journal *Science* selected a Molecule of the Year, a tumor suppressor gene known as p53, the *p* standing for "protein" and 53 being its molecular weight. It seems the level of p53 rises in cells in response to DNA damage that causes mutations that result in cancers. After treatment with gamma radiation, p53 levels can sometimes halt the division of cancerous cells or program some cells to die, a process called apoptosis. Possibly the protein can be used to protect the body against cancer.

On the other hand, a mutant p53 can have the reverse effect, stimulating abnormal cell proliferation and therefore causing cancer. At least half the cancers that afflict humans have this p53 mutation, and they are the most frequently fatal: cancers of the breast, colon, prostate, lung, liver, cervix, skin, and bladder. In addition, several links with environmental carcinogens have been found, including those in tobacco smoke and a type of virus associated with cervical cancer. These factors and others target p53, either by altering one of its amino acids in order to change its beneficent function or by producing proteins that can destroy it. Evidently a deeper understanding of p53's role in cell division may lead to techniques that will eventually stall or prevent the undisciplined cell growth that produces malignant tumors.

This line of research seems very promising at the end of the century, but we would do well to recall how many times throughout the past hundred years the expected line of development has been cut short or diverted or completely overtaken by the unexpected or serendipitous. There is no guaranteed road map for a true odyssey.

Will the retroviruses, like the frightening Ebola virus that causes a horribly painful death, travel from their hidden enclaves in untouched jungles or in animal populations and ravage human populations? Some experts fear so, especially as development levels the forests and brings humans into closer and more frequent contact with infected animals. If carbon dioxide levels in the upper atmosphere continue to rise, causing the greenhouse effect that would warm the earth by only a few degrees of temperature, will disease-carrying insects thrive over larger parts of the earth, bearing new plagues to new venues? Some experts in infectious disease believe so. Or will the odyssey of medicine encounter challenges from a completely expected quarter?

Life will not be fully cleansed of disease and physical malfunction in the near future, perhaps never in the ongoing human story. Most of us will always believe that death comes too abruptly, whether for ourselves or for those we love.

But medical science, if not capable of producing lives eternal and free of pain and disability, is likely to continue making advances that surprise expert and layperson alike, while making our brief sojourns more pleasant and a little lengthier all around. One lesson of the past century in research medicine is that we cannot predict what will be learned, but we can reasonably expect it to be astonishing, life-enhancing, yet soon assimilated as normal, as part of the essential human condition. We soon take any miracle for granted. Our amazement fades. We seek the next cure, the deeper explanation. We are—to our credit—never satisfied.

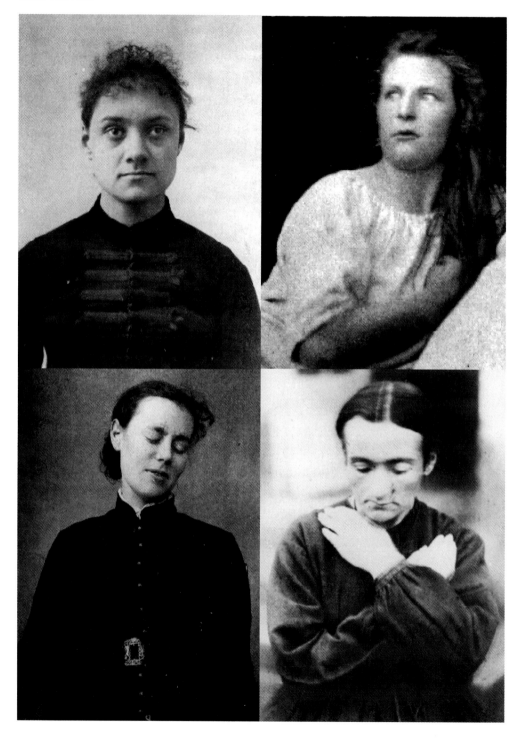

CHAPTER FIVE

IN SEARCH OF OURSELVES

When we began inventing machines to serve us in the nineteenth century, they reflected our own image, weaving cloth and sawing lumber and lifting heavy weights. They had to be fueled. So did we, for everything in the natural world—with no distinction made for humankind—was driven by a force called energy.

So theorized the German physicist Hermann von Helmholtz in 1847, and the idea took hold. It had the appeal of seeming to be self-evident. Just as the machines of the industrial age hammered away, if properly maintained, with all parts working in concert, so did the human body hum efficiently along, a cooperative crew of muscles, nerves, and organs designed and built to produce.

But machines break down. Pushed to extremes, the human body can suffer muscle fatigue or exhaustion of the brain or spine. At the turn of the century it became clear that the emotional results of this exhaustion included depression, insomnia, anxiety, and apathy. Scientists became preoccupied with this problem: How can the human machine be operated with maximum efficiency? Anson Rabinbach has described in *The Human Motor* the "widespread fear that the energy of mind and body was dissipating under the strain of modernity."

An innovative nineteenth-century researcher in this field was Étienne-Jules Marey, a French inventor and physiologist. Using photography and various odd measuring contraptions at his Physiological Station in Paris, he created the very first time-motion studies. For example, he took sequences of photos of human runners and pole vaulters in action—a process he called photochronography—in order to learn exactly

Research into hysteria at the beginning of the century
focused on its outward manifestations as possible clues
to hidden physical causes of the illness.

how we move through time and space. He also studied the movements of horses and sheep.

These photographic records revealed what the eye could not see or the brain understand in real time: exactly how the sprinter moved every part of the body at high speed and how the four hooves of a horse hit and left the ground in succession. Once an action was broken down and studied, Marey believed, it would be possible to improve it. His various devices also recorded motion in the form of curves or lines and measured the pressure exerted by a jumping athlete.

Marey not only illuminated how we move. His discoveries led him and others to study how our movements are fueled—in other words, how much energy we normally have available, how it is spent in activity, and whether or not we can learn how to use less while performing the same amount of work. Marey recognized, as John Hoberman has shown in *Mortal Engines,* "that his own physiological research implied a search for the ultimate capacities of certain organs." The general assumption, building upon Marey's work on fatigue, was that the body draws on a fixed amount of energy.

The effort to capitalize on energy while avoiding fatigue became almost a cultural obsession in developed countries. There was great excitement when a German scientist announced that he had created a fatigue vaccine. It failed, but research continued.

The Nervous Century

Then popular imagination was piqued by a pleasantly exotic theory about mental fatigue. George Beard, a physician in New York City, proposed in the 1860s that all types of fatigue be subsumed under the blanket term "neurasthenia." Running the gamut from mild emotional stress to severe neurosis, but falling just short of raving insanity, this condition was the result of exhaustion of the body's nerve cells. In other words, the mind, just like the body, has a bounded store of energy.

Dr. Beard's idea caught on. Sometimes called nervous exhaustion or nervousness, neurasthenia became a kind of disease of choice as the world faced the twentieth century. Some said that the coming age would become the "Nervous Century," especially in America, where rapid progress in various unsettling forms was jangling the nerves. Never before had life seemed so unrelentingly stressful. When trains attained the unprecedented high speed of 40 mph, would the human body be able to withstand the pressure? One prominent neurologist wrote, "The primary cause of neurasthenia in this country is civilization itself with its railway, telegraph, telephone, and periodical press intensifying in ten thousand ways cerebral activity and worry."

Dr. Beard, as often happens with ambitious healers, had weathered a personal crisis himself. As a young adult he endured a period of stress and nervous confusion. After he recovered, he pursued a career in medicine, and when he detected so many similar problems among his patients, he decided that neurasthenia was a pathology born of his time and place.

The country pretty much agreed with him, and an unprecedented notion became commonplace: that ordinary people undergoing the mental stress of daily life had a medical condition. Was the arrival of the twentieth century too tough on the human race? In an era that was still adjusting to Darwin's theories of natural selection, did the increase in neurasthenia indicate that the human species might not survive modern society?

Nervous disorders, as Francis Gosling has written in *Before Freud*, seemed to be cropping up everywhere. The human machine was definitely not perking steadily along.

Using a special camera he developed in 1881, Étienne-Jules Marey studied human and other animal locomotion by snapping a series of photos that seemed to halt the movement of limbs. In this sequence made in 1892, his so-called chronophotographic apparatus shows the approach, leap, and landing of a jumping man. Although his device was a forerunner of the movie camera, Marey was primarily interested in the effects of fatigue upon human beings.

"The patient may be likened to a bank," said Dr. J. S. Green of Boston, "whose reserve has been dangerously reduced and which must contract its business until its reserve is made good; or to a spendthrift who has squandered his inheritance; or to a merchant who has expanded to the point of bankruptcy."

Many in the general public became alarmed, since traditional science did not immediately offer remedies for this seemingly incurable problem. The layperson's concern was not unlike public reaction toward the close of the century to a similarly inexplicable, similarly culture-derived but no less *real* affliction, the eating disorder known as *anorexia nervosa*. Then as now, a great many people had seen examples of a devastating illness firsthand. And the disease was even more frightening because physicians could offer little effective help.

Inevitably, creative entrepreneurs filled the void for neurasthenia, often to great acclaim. Early ad copy gravely promoted the soft drink Coca-Cola as "the ideal brain tonic for headache and nervousness." The public already believed that drinking the appropriate substances, such as "nerve tonics" or phosphate sodas, would "build up" the wearied nerves. For the specific problem of "overbrainwork," so-called brain salts were

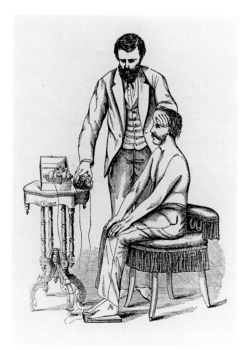

Left: *In this illustration of the principle of "general faradization," the application of a type of electricity to a general area of the body like the spine or head, the operator completes the circuit by touching the treatment area while the patient receives minimal electric current through his bare feet. With his left hand, the operator raises or lowers the current to maximize the effectiveness of the treatment.*
Right: *The search for modern cures for modern illnesses, especially stress, led to the invention and marketing of such questionable devices as the electropathic belt, electric hairbrushes, and other electrified objects in the early twentieth century. The stimulus was in part the work of English scientist Michael Faraday, who had developed the first dynamo the century before, thus showing how the invisible power of electricity could be harnessed for the benefit of humans.*

recommended. Scientists had learned that nerve cells are made up of chemicals. If neurasthenia was the result of chemical depletion, why shouldn't the right chemical solution restore the right balance?

Researchers had discovered that nerve cells operate by sending and receiving electrical impulses. Vulgarize this a bit, and you can see the need for stimulating exhausted (or electrically depleted) nerve cells with outside electrical stimulation. Grand new avenues of treatment opened up: the Electropoise Machine for reviving fatigued nerves (thus curing neurasthenia) and Dr. Scott's Electric Hairbrushes and Electric Corsets.

Dr. Beard himself pioneered the use of low-voltage forms of electrical current in electrotherapy, sending current through paddles, poles, and rods to stimulate nerve fibers and brain cells. His treatments did not fit within any mainstream scientific theory of the day, but he successfully medicalized a condition, from diagnosis to therapy, that had been previously considered on the order of "wear and tear."

The Great Leveler

The rising tide of neurasthenia became a source of national anxiety, and many Americans began to worry about their worries. This was no nuttier than patriotically exulting, as some did, that American neurasthenia proved that the country was more advanced than the benighted, if calmer, world beyond our shores.

In 1904 some physicians warned their afflicted patients to avoid the mental hazard of attending the World's Universal Exposition that year in St. Louis. Said one: "To the neurasthenic, a visit to this greatest of World's Universal Expositions is a risky experience, unless taken under neurologic advice with due experienced psychiatric precautions." The neurasthenic tourist was advised to see no more than one building a day at the 1,240-acre exposition and plan on a total visit of forty to sixty sauntering days.

Originally neurasthenia was a disease associated with the work of professionals, or so-called brainworkers, but it soon embraced all classes . . . and smoothly assumed new guises. When muscleworkers—that is to say, the working class—complained to their physicians about such classic symptoms of neurasthenia as anxiety, insomnia, and depression, they were diagnosed as having the particular form of neurasthenia triggered by physical labor: spinal congestion.

No one was more vulnerable, however, than America's women. In the first place, they were born at a disadvantage since they enjoyed a smaller supply of energy than men and therefore had less capital to squander. Secondly, their nervous systems were thought to be controlled by their reproductive organs, which were very demanding on the energy supply.

Put most succinctly, women were supposedly forced by nature to choose between brains and babies. Too much "brain activity" would deplete the energy required for motherhood; a woman using her brain was risking nervous bankruptcy. Many women were demanding access to higher education, and the American Woman Suffrage Association was clamoring for women's right to vote, but the science of the day—concerned for women's well-being—found the national disease, neurasthenia, to be a natural obstacle to further social change.

The "Woman's Disease"

Meanwhile France did not restrict itself to a single national nervous disorder. At the end of the nineteenth century the distinguished neurologist Jean-Martin Charcot treated a

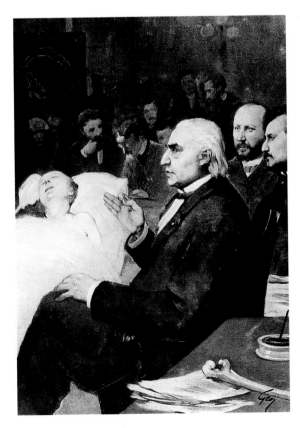

In this painting, Jean-Martin Charcot has evidently hypnotized one of his patients at the Salpêtrière Asylum as part of his lecture-demonstration before a group of invited medical professionals, writers, and other members of the Parisian intelligentsia. Immediately to Charcot's right may be Paul Richer, who made drawings that emphasized the patterns of expression exhibited in sufferers from hysteria. Under hypnosis, this patient would be asked to demonstrate her symptoms.

variety of mental afflictions at the Salpêtrière Asylum, a sixteenth-century arsenal converted into both a woman's hospital and medical laboratory on the outskirts of Paris.

Like most scientists in Europe at the time, he believed in hereditarianism, the notion that the traits of character and personality and morality acquired over a lifetime are passed down from adult to infant. In other words, the child inherits the parents' behaviors, including neurological diseases. These conditions were also believed to change wonderfully as they traveled down the family tree. For example, an epileptic's illness might reappear as multiple sclerosis in his son and, further down the line, as alcoholism in his granddaughter.

Within this framework Charcot assumed that the brains of victims of nervous disorders were weaker than average, a deficiency inherited at home. In each succeeding generation, therefore, the brain would be even weaker. In addition, the damage caused by the stress of modern life could also be passed down like any other factor in producing illness. This combination of weakness and damage would produce increasingly neurasthenic descendants, for whom no cure was possible.

Charcot's analysis brought a new perspective to a nervous disorder considered more serious, more intractable than any other, hysteria. This was a "woman's disease"—

the name comes from the Greek for "uterus"—a disturbing condition that offered a surplus of symptoms from blindness to paralysis, tics to fits and spasms, coughing to feelings of choking to death. The symptoms reflected the position of the uterus, which was thought to wander willy-nilly around the body, causing mayhem. An alternative theory held that hysteria was an indication of sexual promiscuity. But Charcot did not ignore the evidence of his senses. He recognized that men also fell prey to hysteria.

This was a quantum leap in perception. In effect, Charcot moved the seat of the disease from the reproductive organs to the brain. He thus legitimized hysteria as a mental disorder that could be studied, since he assumed that the cause must be a lesion in the brain itself.

When he could find no such lesion in the autopsies of deceased victims, he adopted one of science's most venerable approaches to the unknown: He studied the abundant visible symptoms of hysteria in order to expose the underlying pattern that

Left: *Charcot hired photographer Albert Londe to document the wide gamut of troubled expressions and contorted gestures displayed by his patients.* Below: *One of Richer's drawings focuses attention on the physical symptoms of hysteria by freezing a victim's thrashings in midseizure and deemphasizing her personality.*

might define it. By cataloging the thousand flailing gestures of the hysterics at his asylum, he looked for coherency and pattern, hoping to develop a neurologically based model of some scientific elegance.

"He used to look again and again at the things he did not understand," one of Charcot's students recalled years later, "to deepen his impression of them day by day, till suddenly an understanding of them dawned on him." He came to believe that the physical contortions yielded clues to the nature of the inherited brain damage within.

Artist as well as researcher, Charcot sketched many patients, then eventually hired Albert Londe as director of photography to record their actions. Some ten thousand images remain from his efforts, which he conceived as a "museum of living pathology." On the one hand, convinced that the camera always reveals the truth, he adopted the new technology of scientific photography. On the other, many of his photos look like kin to Renaissance paintings, for he wanted his work to integrate with the European cultural tradition at its best. The odd poses sometimes acquire a haunting beauty.

But Charcot's driving intent was not art but documentation. He hoped to record both the positions that characterized hysteria and the characteristic sequences of those positions. Gradually, as Sander Gilman has pointed out in *Seeing the Insane,* the photos lost favor as research tools: The camera caught actions so specific to each individual subject that it was difficult to see patterns that would apply to a large group of patients. Instead, the composite etchings of Paul Richer, whose work combined the actions of several patients into one image, made possible the illustration of various distinct stages of hysteria shared by the afflicted.

An inspired performer, Charcot also gave weekly lecture-demonstrations that attracted visiting physicians from around the world as well as the artistic and intellectual elite of Paris. Typically the all-male audience watched solemnly as he hypnotized one of his female patients, then guided her in demonstrating the essential symptoms of the disease. This scene became so widely known that every circus in France boasted a hypnotist in its sideshow, always billed as "le docteur Charcot."

The Anatomy of Error

One student at the Salpêtrière was a young Viennese neurologist who worked on a fellowship for four months in 1885 and 1886. Sigmund Freud was impressed with the older man's clinical skills, but he had learned too much in neuroanatomy at other European laboratories to accept Charcot's theories about hysteria. He had studied the brain struc-

Shown with his wife, Martha, the solidly bourgeois Sigmund Freud first argued in his 1900 work The Interpretation of Dreams *that the hysterical symptoms studied at the Salpêtrière indicated repressed but powerful sexual conflicts that had been produced by experiences in childhood. Unlike some of his followers, Freud recognized that his "talking cure" would benefit neurotics but not severely disturbed schizophrenics.*

tures of fish, eels, and human beings. Because he was familiar with the breakthrough work of physiologist Ernst von Brücke and Theodor Meynert, neurologists who had delved into the structure of nerve cells at the microscopic level, Freud understood a great deal about the anatomy of the brain, including such astonishing new discoveries as the location of the centers that control vision and movement.

To this point Freud had believed that brain functioning could be understood by applying the principles of the physical sciences. Watching the victims of hysteria, however, he gradually began to suspect that the cause could not possibly be physical. "Hysteria behaves as though anatomy does not exist," he concluded, "or as though it had no knowledge of it."

For example, one apparently paralyzed patient let her hip drag along as lifelessly as her leg. But according to basic anatomical principles, the hip would not have been affected by a physical paralysis of the leg. Freud was also intrigued by so-called glove anesthesia, in which numbness extends from the hand to the upper arm. Again, a patient defied the rules of anatomy: He let his hand hang limply at the wrist, as if the numbness stopped short there.

Freud also knew that Charcot's numerous autopsies had failed to find lesions in the brains of hysterics. Now that the physiological avenue of reasoning failed him, Freud began to look elsewhere for explanations of hysteria as well as other emotional conditions.

Eventually his probes into the unconscious became a rejection of the day's orthodox hereditarian views, which had ordained that Jews and certain other groups were especially susceptible to nervous degeneration. Jewish himself, with a clientele that was more than half Jewish, Freud had no patience with what he sensed was an anti-Semitic theory of mental distress.

"Nervous" Anna O.

When Freud returned to Vienna and opened his private practice, the distinctive characteristics of his clientele helped determine the nature of his psychological discoveries.

The psychiatrist Josef Breuer, pictured here with his wife, was one of the young Freud's mentors. It was he who actually treated a neurotic famously associated with Freud's theories, Bertha Pappenheim. Breuer wrote that under hypnosis he "relieved her of the whole stock of imaginative products which she had accumulated" since his previous session with her.

Most of the patients at his clinic at 19 Berggasse were attractive young women from comfortable or even opulent backgrounds. If they exhibited bizarre behavior, they also tended to be beguiling, easing the young man's move from brain sectioning and other pale operations of the inanimate laboratory.

Bertha Pappenheim, called Anna O. in Freud's work to protect her privacy, could have fitted well with these distressed, sometimes distressing females, but this famous inspiration for his theories never actually met with him. A luminously bright twenty-one-year-old from an Orthodox Jewish family, she was under the care of neurologist Josef Breuer, one of Freud's mentors. Pappenheim had been considered "nervous" from her

earliest years because of a penchant for daydreaming. What might have been considered eccentricity within the domestic circle became much more disturbing after she nursed her father through a severe illness. She lost sensation in an arm and leg and became paralyzed. She could not bear to drink water. Always gloriously articulate, she forgot how to speak.

Ironically, it was a return to speaking, or what she treasured as "the talking cure," that alleviated her eerie symptoms, but only with expert help. Every day she met with Breuer to be hypnotized, and only in that state could she consciously recall, relive, and discuss a series of forgotten personal traumas. Eventually the vise of her inner experience relaxed; the physical manifestations of her "hysteria" disappeared.

Breuer told Freud it was probably the act of talking, not the hypnotic state, that was effectively therapeutic. The younger man was struck by the specificity of Pappenheim's symptoms and began to speculate, drawing upon what he had seen at the Salpêtrière and within his own office. The young woman's symptoms—the paralysis, the inability to speak—were not random and meaningless. Nor were they significant by themselves. Rather, they were a kind of road map, a private code, that might provide clues to the nature of a deeply troubled, deeply hidden inner life.

In short, Freud decided that the patient's concealed story was the answer to hysterical dysfunction. If the doctor could elicit this key, if he listened as the story was told, healing would begin.

Histories Within

But how to find the story? Hypnosis had not proved a reliable gateway. Freud began placing his hand on the foreheads of his patients—there were Anna O.'s in abundance for him to work with—and encouraging them to ramble aloud through their lives and worries. This technique later became celebrated as free association.

While the patient was blithely free-associating, young Freud was concentrating with the ferocity that we associate with genius. In these ramblings, as he turned his own perceptions back and forth and around, he came to a conclusion that founded a new

In his study in Vienna, Freud relaxed on the chair and footstool to the left, out of the line of sight of a patient reclining on the couch. Thus his presence would be less likely to hinder the process of free association, or the apparently unstructured flow of memories, ideas, and thoughts as they came to the patient's consciousness. When a particularly important revelation or connection was made, Freud was likely to celebrate, as he termed it, by rising to smoke a cigar.

discipline of medicine and fathered a billion silly remarks at cocktail parties and in romantic situations for decades to come.

This so-called hysteria, Freud decided, sprang in its peculiar violence from the experiences that lodge most irrevocably in the memory. All of us, not just the interesting young women of bourgeois Vienna, deal with unbearably painful experiences—as well as the milder unresolved conflicts of daily life—by swiftly banishing them to a mental cooler Freud called the unconscious. In other words, we repress them, but they are not so easily dismissed.

Later the repressed conflict reemerges, not as thought or verbal expression but as physical symptom indicative of the trauma. Pappenheim's paralysis in selected limbs and her hydrophobia were concrete signs of specific types of inner emotional turmoil. Freud's stunning, radical break from traditional medical practice, as well as from traditional ideas about the relationship between mind and body, reads something like this: The language of the body discloses the emotional history of the mind.

When Charcot saw a hysteric's panoply of physical symptoms, he looked to family history and was likely to find a pattern of intensifying degeneration. When Freud saw the same twitches or paralysis or diminished capacity, he listened. Perhaps more important, he thought there was the possibility of comfort or even cure. As the philosopher Michel Foucault put it seven decades later, Freud and his followers believed in "the possibility of a dialogue with unreason."

This original approach toward understanding the unconscious developed from the physician's need to heal himself. Although he was happily married and deeply attached to his growing brood of children, Freud worried about his health, about his mortality. These fears only heightened when his father died, a loss that hit him very hard. But in the depths of his despair he came to realize that his dreams offered signal clues to the exact nature of his draining subconscious fears. He not only emerged from this dark period with a feeling of liberation but also inspired a worldwide movement in therapy, including the analysis of dreams.

Not everyone immediately accepted his ideas, of course. In particular, critics were offended by his controversial ideas about human sexuality. In the simplest terms he argued that every human conflict is basically a sexual conflict. Moreover, he believed that sexual feelings are present from birth onward, an idea that horrified many who did not want to think of children as sexually aware beings with sexual desires and fantasies that became engrained at an early age in the unconscious. Finally, Freud's insights had not been, nor could they be, confirmed in the laboratory or otherwise backed up with material proof.

In due course he invited several followers, or perhaps "devotees," to meet weekly with him to discuss his new concept, psychoanalysis. Many were separated from the mainstream of Viennese society. They zealously took up Freud's ideas as a kind of faith. This unfortunate aura of religiosity did not encourage scientists to ease their skepticism.

But Freud's theories about the unconscious, and about the inner conflicts that rage unseen within the individual, were soon radically affirmed by traumas experienced in a conflict that engaged almost the entire world.

Nerve Warfare

I heard the most terrifying thing I ever heard in my life—the loud, malicious scream of a big shell. . . . I heard a man screaming, holding his body with both hands. . . . Then my eyes rested on a boy who had laughed . . . blood was pumping out of his body like red water. . . . My head hurt, my face hurt, my eyes and ears. . . . I hurt all over.

As trench warfare reigned over the front lines of Europe, soldiers told war stories like none ever heard before. The Germans first saw this strange and terrible phenomenon. Their Belgian and French adversaries, driven into the open by the massed shelling, stumbled about "with staring eyes, violent tremors, a look of terror." Many wept uncontrollably.

At first some Germans took these scenes as evidence that the enemy was a constitutionally weak, psychopathic, degenerate race. But when the armaments of the Allies began to equal theirs in power and number, their own men began to display the same baffling symptoms.

Since bombardments seemed to be the proximate cause, the condition was called shell shock, but to name was not to explain. Besides, there were other odd happenings. After an explosion some of the dead had no visible wounds, and autopsies revealed only tiny hemorrhages in the brain and spinal cord. Were these deaths, and indeed the fractured behaviors of survivors, caused when the concussion of an exploding shell minutely damaged nervous tissue?

This premise was itself exploded when soldiers far from action began to have fits or suddenly become completely deaf or mute—in other words, exhibit so-called shell shock. Some hard-line commanders had no patience with this mysterious, apparently inexplicable malady. Their suggested therapy was summary execution for cowardice, and many of the afflicted were indeed tried by military courts and shot. But punishing every last victim of the disorder would have required killing eighty thousand soldiers.

As physicians tried isolation, rest cures, and electrical prods with little success, some began to realize that "shell shock," "soldier's heart," and "traumatic war neurosis" were—though it was almost shameful to think so—just so many terms for male hysteria. The manly pursuit of war had produced its own raw battalions of hysterics.

Freud's work suggested an answer. The soldier in this vile new system of trench warfare was faced with an impossible situation: He could not choose either to fight or to flee—the comprehensible options of traditional battle—but could only sit dumbly still and wait out his chances of death. Shell shock was the body language of powerlessness.

As one soldier recalled, "It was simply a case of looking death in the face and waiting to be hit. I never got into a worse hell." As millions of young soldiers endured this peculiarly twentieth-century form of torture, the fight or flee conflict was internalized, and then the men began exhibiting the symptoms of hysteria. This analysis brought understanding and broke a pathway toward relief. A British neurophysician summed up the dilemma: "From the combatant's point of view, the Great War has been described as industrial warfare; from the medical point of view it might be described as nerve warfare."

Partly because "shell shock" became a household term, Freud's theories began to flow from the drawing rooms of Vienna into the mainstream. Even if psychoanalysts and their treatments were not entirely trusted or respected by all, virtually everyone in the general public began to accept the idea of psychologizing behavior—in other words, to believe that the mind and its secrets explain why we act as we do and that environmental factors play an important role in the development of our inner lives.

Morons in America

But there were alternate explanations, and they had an enormous impact upon public policy in America. If hidebound hereditarianism had been discounted by Freud, there were still scientific researchers who saw heredity as a strong factor in human behavior.

The United States at war's end, in sharp contrast with both its Allies and its enemies, was a land unscarred by conflict and eager to enter the golden age of technological progress. Then came a huge shock: The country that had never been attacked on its own soil might well be defeated from within. If Army intelligence tests released after the war were accurate, the average mental age of native-born U.S. soldiers was thirteen years old; indeed nearly half the whites were virtual morons. Other races ranked even lower.

Did this reflect a larger truth about the entire society? The intelligence tests in question had been developed by prominent psychologists—Lewis Terman, Robert Yerkes, and Henry Goddard—as a system for quickly matching some two million recruits with specific military tasks as well as weeding out misfits. Terman explained, "If the Army machine is to work smoothly and efficiently, it is as important to fit the job to the man as to fit the ammunition to the gun." The three men had previously used intelligence testing in their private practices, but only on a one-on-one basis. Volunteering their expertise for the war effort, they came up with the first standardized intelligence test ever designed to be taken simultaneously by hundreds of people within less than an hour.

Actually there were two versions: Alpha for recruits literate in English, Beta for everyone else. On the former there were the kinds of multiple-choice questions that have since become gratingly familiar to succeeding generations:

Pictured around 1916 in one of the horrific trenches of the European conflict, these British troops were typical of the young men exposed to the stresses that caused "shell shock." For some time, military professionals resisted the psychological implications—i.e., that men profoundly traumatized by conflicting emotions under fire could exhibit the symptoms of hysteria thought peculiar to women. Many of the afflicted were punished, even executed, for presumed cowardice.

The most prominent industry of Gloucester is

  A. *Fishing*   C. *Brewing*

  B. *Packing*   D. *Automobiles*

                    or

Christy Mathewson is famous as a

  A. *Writer*   C. *Baseball player*

  B. *Artist*   D. *Comedian*

The Beta tests are less widely known. Young men who were recent immigrants to the United States or had been reared in poor rural areas without access to radios or illustrated newspapers were asked to complete drawings that lacked essential visual information. Not surprisingly, they might fail to guess how to fill in a half-erased picture of an urban American game like bowling or tennis.

In both tests, obviously, the presumed yardstick of native intelligence relied primarily on native cultural knowledge; the number of pins at the end of a bowling lane or

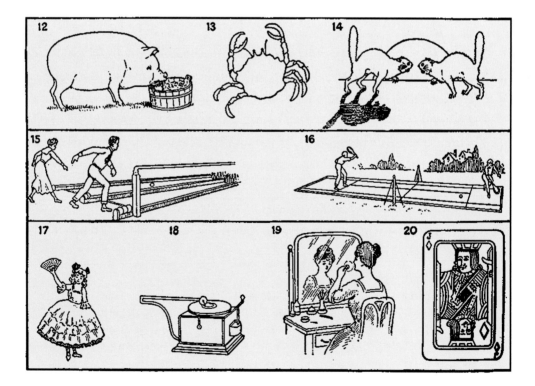

the principal occupation of a small Massachusetts town are not inherited items of information. Yet the Army psychologists believed that their Alphas and Betas accurately calibrated innate intelligence. Therefore many in American government assumed that urbanization, which had been fueled both by country folk looking for jobs and by refugees from abroad—more than twenty-three million of the latter between 1880 and 1924—was producing stewpots of disease, ingrained stupidity, immoral behavior, and crime. The average level of intelligence was being drastically lowered by the increasing number of rubes and aliens crowding into the cities.

Earlier in the century experts had believed that all these problems resulted from "feeblemindedness," a catchall term that covered a wide spectrum of vaguely defined deficiencies. At Ellis Island, where potential immigrants were cursorily assessed by a harried, understaffed medical team, obvious or detectable physical illness was only one justification for exclusion. Immigration officials were most deeply worried that they would unknowingly admit the feebleminded. They did not have sophisticated diagnostic tools or medical technology for analyzing mental problems.

In practice the diagnostic tool at Ellis Island was the individual physician's intuition. He might determine that a "strange" way of dressing—perhaps the village garb of a poor immigrant from deep within Eastern Europe—revealed a feeble mind. By regarding the facial expression of someone who might well be apprehensive, overeager, or confused in the medical cattle call at the gateway to American opportunity, he might diagnose melancholia or drug addiction from a tic or glance or flinch.

The efficacy of these techniques was not strengthened by the conditions. On April 17, 1907, the busiest day in the island's history, 11,747 immigrants were processed. Medical observers called over anyone limping and wrote *L* for "lame" on his or her clothing with chalk, *H* for "heart problems" if anyone seemed out of breath walking up stairs. On average, 1 in 50 potential immigrants was rejected as medically or mentally unfit to enter America. Approximately 3,000 of them killed themselves rather than be sent back.

A page from the test used to assess the intelligence of army recruits in World War I reveals some of the unconsciously prejudicial assumptions of the testmakers. The sons of recent immigrants or dirt farmers would be unlikely to understand the rules of the sports of their more leisured fellows. The results of such tests alarmed anyone who assumed that native intelligence is best measured by one's familiarity with middle-class American life.

A city health officer examines immigrant children upon their arrival at Ellis Island off Manhattan in 1911. At the time, there was particular concern about the importation of typhus, which had reached epidemic proportions in eastern Europe. Legitimate concern about potential epidemics, however, was not always the basis for rejecting hopeful immigrants, whose unfamiliar dress or behavior was sometimes assessed as an indication of mental aberration.

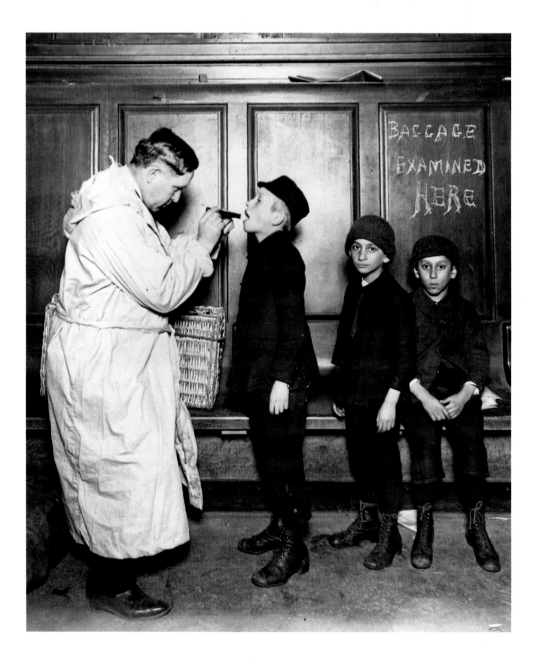

No one was pleased with this situation. It seemed prudent to seek the assistance of volunteer Army psychologist Henry Goddard, the nation's leading expert on feeblemindness and director of the Vineland Training School for Feeble Minded Boys and Girls in New Jersey. His concentration on psychoasthenia, or weak minds, paralleled the nation's earlier worries about neurasthenia, or weak nerves.

It was Goddard who introduced the Binet intelligence test to the United States after the war. Developed in France by Alfred Binet, who hoped to cull out mentally retarded children in order to give them special help, it became regarded as a tool for sorting out all children for classroom placement in accordance with their supposed native mental abilities.

This would seem to be a beneficent exercise. Surely children are best served when the teacher teaches them at the appropriate level, and adults can make happier career choices when they find out where their strongest skills lie. But there is a potential dark side to tests that can be used to exclude rather than include, as Goddard unwittingly demonstrated.

Not coincidentally, he was the inventor of the term "moron," which he defined as any adult with a mental age of between eight and twelve. As he traveled back and forth to crowded Ellis Island, purveying the Binets, he gradually assembled data supposedly showing the percentages of feeblemindedness evident in immigrants by race: 79 percent of Italians, 80 percent of Hungarians, 83 percent of Jews, and 87 percent of Russians.

Such results could not go unchallenged, of course, but psychologist Goddard was convinced: He believed in his methods. Overall, something like three quarters of all prospective immigrants must be morons.

The Kallikak Breed

Goddard worked to find the cause of the phenomenon he had discovered. Choosing heredity as his point of departure, he firmed up two assumptions: The individual's feeblemindedness (1) is both congenital and permanent and (2) will inevitably be passed down to the next generation.

For his theories, no evidence was more compelling than the bizarre history of the Kallikak family. Goddard portrayed them as successive generations of the mentally defective, criminally inclined, poverty-stricken offspring of a casual liaison between Revolutionary War soldier Martin Kallikak and a mentally defective young woman he met in a tavern. Of the 480 descendants of their son, Martin Jr., Goddard reported that

"143 were or are feebleminded, while only 46 have been found normal. The rest are unknown or doubtful."

Moreover, the primal Kallikak eventually married a woman of high moral character, and their progeny were apparently equally upright through the corridors of time. Goddard therefore concluded that the woman of the tavern episode was alone responsible for the feeblemindedness of the more colorful Kallikaks. He had his answer: "No amount of education or good environment can change a feebleminded individual into a normal one, any more than it can change a red haired stock into a black haired stock."

Unfortunately both the tale of the bad Kallikaks and Goddard's theory became ingrained in popular culture. But later experts proved Goddard wrong. The very concept of "feeblemindedness," which was used for a grab bag of so-called moral character

Psychologist Henry Goddard determined that the descendants of a liaison between Revolutionary War soldier Martin Kallikak and a "feebleminded" barmaid tended to be "degenerate," as shown on the right, while his marriage to an upstanding bourgeoise with full mental capacities produced generations of physically and mentally healthy progeny successful in the arts and academic life.

traits, is too broad to be defined as a simple genetic trait. Despite the apparatus of "scientific" tests, this was a subjective analysis of subjective concepts.

"The Better Elements"

Yet the concept of biology as destiny had insidious appeal. It nourished a movement called eugenics, which aimed to "improve" the human stock by applying the principles of heredity. To scientists and social reformers alike, to U.S. presidents and university presidents, and even to well-meaning celebrities of achievement like Helen Keller and Alexander Graham Bell, eugenics seemed to be the pathway to a better society. In 1924, pressed by eugenicists, President Calvin Coolidge signed an immigration bill that helped stave off "race suicide" by limiting the dark-featured influx from southern and eastern Europe.

Eugenicists also fervently advocated applying the principles of raising farm animals to the breeding of the optimum American family. The American Eugenics Society explained this reasoning in a brochure it distributed widely: "The time has come when the science of human husbandry must be developed, based on the principles now followed by scientific agriculture, if the better elements of our civilization are to dominate or even survive."

At state fairs around the country Fitter Family Contests and Better Baby Competitions sponsored by the society required entrants to pass both physical examinations and intelligence tests. Winners and losers alike might profit from exhibits that somberly echoed Goddard's theories, as in this poster at the 1929 Kansas Free Fair: "Unfit human traits such as feeblemindedness, epilepsy, criminality, insanity, alcoholism, pauperism, and many others run in families and are inherited in exactly the same way as color in guinea pigs."

There was also a third way to encourage the "better elements," in addition to immigration bars and human husbandry: In 1914 the American Breeders Association proposed that "defective classes be eliminated from the human stock through sterilization." In addition to the ambiguously designated "feebleminded," the defective were thought to include paupers, criminals, epileptics, and the insane.

Such thinking had long been in the air. Before the Great War more than a dozen U.S. states had specifically passed sterilization laws, and almost all had assumed the power to sterilize by force the insane, criminals, and epileptics held in institutions. By war's end, however, most state governments had backed off, uncertain of the constitutional issues that might be involved.

To reverse this unwelcome trend, activist eugenicists worked to bring a landmark case before the Supreme Court. In 1924 they got their wish in the matter of seventeen-year-old Carrie Buck, who with her mother and infant daughter was institutionalized at the State Colony for Epileptics and Feeble-minded in Lynchburg, Virginia.

On the Binet test Carrie scored a mental age of nine years; her mother fared worse. After some tests on Carrie's baby girl and a Red Cross worker's testimony that she did not "look normal," Chief Justice Oliver Wendell Holmes ordered Carrie to be sterilized in a sourly memorable phrase: "Three generations of imbeciles are enough." Eugenicists rejoiced for the future of the nation, and twenty-four states passed sterilization laws before the end of the decade.

In Europe there were similar programs to allay the general public's fears about the proliferation of "degenerates." Especially in Germany, as articulated by Chancellor

Adolf Hitler, there was anxiety about potential pollution of the purer races. To preserve the Teutonic ideal, the Eugenic Sterilization Law was passed in 1933: It provided compulsory sterilization for the supposedly hereditary disabilities of feeblemindedness, schizophrenia, blindness, epilepsy, and severe physical deformity. An official of the Third Reich explained the rationale: "We want to prevent . . . poisoning the entire bloodstream of the race. . . . Therein lies the high ethical value and justification of the law." Over the following three years some 225,000 "degenerates" were sterilized. Without evident irony, the German eugenicists expressed their gratitude for the precedents set a decade before by American eugenicists. The generations of the blind and the epileptic, the physically deformed and the schizophrenic would be scientifically terminated for the common good.

Nurture, Not Nature

Other types of analysis of human behavior and development were vying with eugenics in America. In 1925 psychologist John Watson's book *Behaviorism* was hailed by one reviewer as "the most important book ever written." Like others in the so-called behaviorist school, Watson argued that "there is no such thing as an inheritance of capacity, talent, temperament, mental constitution, and characteristics." Rather, environment plays the central role in shaping our behavior. Therefore society can and should use scientific principles to shape and control human destiny.

The first glimmerings of this revolutionary new approach had appeared in the animal laboratories of the Russian physiologist Ivan Pavlov, whose work Watson had publicized within the U.S. scientific community as early as 1913. As Pavlov was studying the digestive systems of dogs, he happened upon a tantalizing discovery. A dog's natural response to the sight or smell of food is to salivate. But when Pavlov trained his animals to associate their feeding with the ringing of a bell, salivation began with the very first ring, even when no food appeared. In other words, the dogs were "conditioned," in the language of behaviorism, to respond to a certain stimulus.

Could human behavior be determined in the same way? Could systems of rewards and punishments teach or modify ways of behaving? Watson certainly thought so:

At the end of the twentieth century with its history of racial conflict and deliberate genocide, this photograph of a family awarded Honorable Mention, Large-Family Class, as part of a eugenics exhibit at a Kansas fair in 1923 needs no comment. It is nonetheless deserving of study and reflection. "Only healthy seed must be sown!" read a poster popular with the American eugenics movement at the time. "Check the spread of hereditary disease and unfitness by eugenics."

Give me a dozen healthy, well-formed children, and my own specified world to bring them up in, and I'll guarantee to take anyone at random and train him to become any type of specialist I might select—a doctor, lawyer, artist, merchant-chief, and yes, even into a beggar-man and thief, regardless of his talents, penchants, tendencies, abilities, vocations, and race of his ancestors.

Along with his colleague Rosalie Rayner, Watson decided to try his theories on Albert, a happy, fearless infant. From age eight months and twenty-six days through eleven months and fifteen days, the unfortunate Albert was subjected to loud stimuli intended to induce fear of any white hairy or furry object. As shown in film footage of the experiments, the boy at first looked or reached eagerly toward these objects, then flinched at the loud sound. Soon he burst into tears and howled at the very sight of a white furry object. Unfortunately the second stage of the experiments—removing the fear through the same kind of conditioning that had instilled it—was never put to trial. Albert's family moved away. Later behaviorists proved that fears conditioned in children could indeed be successfully removed, but Albert's fate is lost to history.

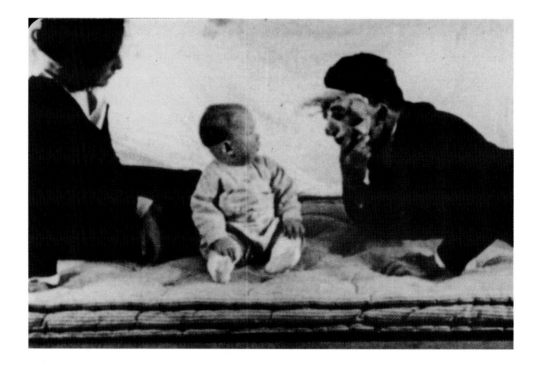

Behaviorism became orthodox in U.S. academic life for decades. Watson's celebrated follower B. F. Skinner became a virtuoso of complex animal experiments. He trained pigeons to play the piano and rats to pull chains to retrieve a marble, carry it to a standing tube, and then drop it in. According to Skinner's theory of operant conditioning, such endeavors were taught by providing an immediate reward for desired behavior or a slight punishment for undesired behavior.

Did these animals "understand" what they were doing? No, but neither, according to behaviorists, do humans do what we do because of our inner mental processes. In other words, consciousness and even Freud's unconscious are therefore irrelevant to our actions in the real world. We do not really know why we carry and deposit the marbles of our daily lives. "We're always controlled, and we're always manipulated," Skinner insisted.

Because the mind is in effect a closed "black box" that cannot be explored scientifically, psychologists can therefore profitably concern themselves only with the external causes and observable results of behavior.

Blind Obedience

Meanwhile the so-called social psychologists, troubled by the ability of dictators in the 1930s and 1940s to transform rational, educated people into blindly obedient masses, began to explore how and why we behave as we do in group situations.

How do leaders affect their followers? How does a group affect the behavior of its individual members?

As the behaviorists continued to educate their lower animals in laboratories, a refugee from Nazi Germany, Kurt Lewin, led a U.S. team of researchers to seek answers to these questions in a radically new kind of experiment that involved the behaviors of human beings. In 1939 he enrolled schoolboys in a study of the contrasting effects of different leadership styles. Three after-school clubs were set up on the Boy Scout model, with various activities that required attention, cooperation, and skills. Three adult males were trained as leaders. One learned to be completely autocratic, making all decisions

Behaviorist John Watson (left) observes an apparently lively, cheerful eleven-month-old "Little Albert." In a series of conditioning experiments published in 1920, Watson and his colleague Rosalie Rayner argued that emotions could be learned. Since a loud noise occurred whenever the inquisitive baby reached for a white furry object, Albert soon burst into tears whenever he glimpsed anything with those characteristics, including Santa Claus.

for the group of boys under his leadership. A second was taught to be laissez-faire, giving almost no guidance to the boys. The third was to take the democratic approach, actively encouraging and assisting the boys in making group decisions. At the end of six weeks each leader was switched to another group, and then a second switch was made after six weeks. During the experiment, in other words, each group of boys was exposed to each of the very different leadership styles.

The results were consistent—and dramatic. Under the autocrat each group worked harder than the others—but only when he was watching. Their docile submission to the adult was matched by increased aggression and hostility toward one another. That is to say, they became little fascists.

Under the laissez-faire regime it became clear that total freedom probably produces chaos. The boys did less than for the other two leaders, and their work was of poor quality.

The democratic approach produced the highest levels of motivation and originality in a group's work and play. In addition, the boys were more playful and praised one another more frequently during their six weeks under the leader who used democratic principles.

The experiment not only vindicated democracy but showed social psychologists that the style of leadership, not the individual leader's personality, and the social situation created by that style are critical to determining human behavior. Whether the worst or the best in human nature surfaces, in other words, depends upon the environment. We act responsibly or ignobly depending upon our situation.

Blind Obedience at Yale

A couple of decades later, in one of the most notorious experiments ever conducted by a social psychologist in the United States, the worst in human nature was easily brought out.

Working at Yale University in the 1960s, Stanley Milgram decided to test whether or not his subjects could be persuaded to administer electric shocks to someone locked in a room when instructed to do so by a supposed authority figure. Each subject was told to push a button that gave increasingly painful shocks to a "learner" whenever he gave wrong answers on a memory test. In fact, no shocks were actually administered, but the equipment produced a credible zap and the learner screamed ever more desperately as the experiment went on.

Beforehand Milgram asked forty experts to predict how far the subjects would go. The majority believed that very few would administer a shock of more than 150 volts, which is moderately painful. Only one in a thousand, they estimated, would administer the upper limit of 450 volts since the instructions clearly explained that such a shock is so dangerous as to be potentially fatal to anyone with a weak heart. Moreover, the supposed scientific expert leading the experiment, actually a high school teacher whose assumed air of authority was accentuated by a white lab coat, told each subject that the learner suffered from a heart condition.

The experts were wrong. Two thirds of the subjects, despite the warning and the piercing cries of the learner, revved up to 450 volts. Thus they chose between the fear of harming another human being and the fear of disobeying orders. Milgram, unsure that anyone would believe his findings, made a film the final day that was first publicly shown in 1997. As a subject administers a shock, a very convincing scream is heard off camera; then the learner shouts again and again, "Let me out of here! My heart's bothering me!" The authority figure orders the subject to continue, and he does. Then a shock is followed by silence. Yet the subject continues following the instructions to push the button. At this point, one subject, in conflict, is shown refusing at last to go on. Afterward, 84 percent of the participants, who included social workers, laborers, and executives, reported that they had enjoyed taking part.

Later many observers resisted the apparent lessons of the experiment, but social psychology seems agreed that blind obedience is part of the basic human condition, an unpalatable characteristic that can be brought out in a controlled situation.

For many people there did not seem to be much good news in the theories of social psychologists, behaviorists, or Freudians. Whether our behavior was determined by a group, by the environment, or by traumas experienced in childhood, the patterns all suggested passivity controlled by outside forces.

None of these avenues of study seemed to hold a reassuring answer to the question that was most pressing at mid-century: How could human beings have committed the atrocities that killed tens of millions before and during World War II?

Becky and Lucy

Not every atrocity had yet made the headlines. During the war young men who had refused to fight because of their pacifist beliefs were deemed conscientious objectors and assigned to noncombat duties. Many were stunned by the conditions they discovered when they worked in state-run mental hospitals.

As if in imitation of the Nazi concentration camps, hundreds of naked mental patients might be herded into barnlike, filth-infested wards. In some institutions rickety cots were crowded together so closely that the floor was completely invisible; in others the mentally ill were confined to bare rooms with no beds whatever. Dim light came through half-inch holes in steel-plated windows by day, but at night the wards became pitch-dark tombs.

In such situations physicians were little more than custodians, perhaps dealing with a doctor/patient ratio as preposterously high as six hundred to one. Treatment, if so it could be termed, tended to be isolation, sedation, or the use of so-called restraints: locks, straps, confining sheets, muffs and mitts, canvas camisoles, thick leather handcuffs. These institutions reeked of hopelessness.

One conscientious objector sneaked a camera into a hospital. When *Life* magazine published his grim images, there was national outrage. Why couldn't psychologists develop effective treatments for the suffering mentally ill, instead of continuing to fine-tune their sweet theories in animal laboratories?

In fact significant treatment progress had begun back in the 1930s in the chimpanzee laboratories at Yale under the supervision of Carlyle Jacobsen and John Fulton. Two chimpanzees, Becky and Lucy, were studied after operations that destroyed a large section of their higher brains known as the frontal lobes. The aim was to analyze how the animals dealt with this measured destruction, which diminished the ability to solve problems.

But Becky's post-op behavior was unexpectedly dramatic on the emotional level. Before the lobotomy she had thrown temper tantrums when she could not perform well on tests; afterward such frustrations left her cheerful. Jacobsen and Fulton assumed that there was some connection between destruction of the frontal lobes and calmed behavior. Jacobsen noted wryly that Becky had joined a "happiness cult." By contrast, Lucy became more prone to anger, as did all other animals after the surgery.

Nonetheless, Becky's transformation piqued the imagination of Egas Moniz, a Portuguese neurologist, who already believed that brain pathology was the basis of all psychiatric problems. In the late 1930s he determined to improve human behaviors by cutting out specific parts of the brain. He thought that the mentally ill have brain cells that become "fixed": These groups of cells, he wrote, "constitute the path of their psychic life; nobody can divert the course of thought of these patients."

His reasoning went something like this: First, we can hypothesize that the frontal lobes are the site of thoughts and ideas; second, we can stipulate that mental illness is caused by "fixed" thoughts that interfere with the flow of healthy mental life; third, we can surmise that these damaging thoughts are lodged in specific neurological pathways. Surgically remove the lobes, and the mind works well again.

Moniz was not following the traditional scientific method of reasoning based upon experimentation, nor was he distracted by the predominant views of the brain in his day, for they inconveniently provided little support. It is also possible that after being short-listed for the Nobel Prize twice, he yearned too keenly for public acclaim. From 1936 onward he began surgically invading the brains of mentally ill patients, using an in-

Immediately after World War II, many institutions for the mentally ill were grisly warehouses of the neglected and forgotten. The worst result of the asylum movement, which had begun idealistically in England in the mid-nineteenth century, was insensitive or even brutal treatment, often coupled with improper or illegal institutionalization. Soon, psychiatrists learned that some mental disturbances were caused or exacerbated by the conditions of confinement.

strument like an apple corer that could be pushed through holes drilled into the skull. In this fashion, connections in the prefrontal lobes would be severed.

In America another neurologist was also convinced that psychosurgery, as it came to be called, was the solution to mental illness. Walter Freeman's own brain had not been surgically altered, but he did ascribe his recovery from a nervous disorder to neurological improvement—in his case produced by vigorous physical activity. He sent off to Paris for his own set of the instruments used by Moniz. Then he and his colleague James Watts began practicing on human brains obtainable from a nearby morgue.

After a week of these dry runs they were ready to operate on a living human being. Almost overnight their work was widely praised here and abroad, and the popular press found many patients happy to express gratitude for the new technique.

PSYCHOSURGERY CURED ME, read one newspaper headline. NO WORSE THAN REMOVING TOOTH headed another interview with a patient. Taking the larger view, one reporter's survey was billed thus: WIZARDRY OF SURGERY RESTORES SANITY TO FIFTY RAVING MANIACS. One severely disturbed woman was able not only to leave the hospital but also to win a championship in whist.

In this somewhat frenzied climate Dr. Freeman warmed to the spotlight and toured the nation tirelessly, even appearing on early television and making demonstration films to "sell" the idea of psychosurgical intervention as a cure for psychotic patients. Indeed such patients, their families, and their caregivers felt justifiably hopeful for the first time in medical history. Even schizophrenics, the most mysterious and intractable of the mentally afflicted, seemed to respond. In 1949, after thousands of patients had been lobotomized, Egas Moniz was finally awarded the Nobel Prize.

But Freeman, like many other experts, did not agree with Moniz's explanation for the success of lobotomies. Indeed he rejected the notion of fixed thoughts entirely and turned his attention toward the section of the brain known as the thalamus. There, he theorized, lay the mind's emotions, while pure rationality was seated in the prefrontal region. In other words, the "emotional" and "thinking" brains were connected, but a lo-

American neurologist Walter Freeman, an enthusiastic popularizer of lobotomies, operates on a patient in one of his many public demonstrations. Thought to alleviate clinical depression or calm the anger and violent actions of the seriously disturbed, the procedure involved severing the nerve connections between the brain's two frontal lobes. Freeman himself designed the instrument he is sliding beneath the patient's upper eyelid to enter the brain.

botomy severed that connection. Mental disorders were cured by effecting a kind of normative balance—that is, emotional reactions became less intellectualized, and intellectual reactions became less emotionalized. Freeman's was the prevailing rationale for psychosurgery throughout the 1940s.

But time eroded public and expert confidence in the miracle cure. Many patients showed little or no improvement until a second or third operation. Many showed no improvement at all or became even more disturbed. To some critics the procedure was simply too drastic on its face, a spooky intrusion into precious organic material that was not fully understood. Eventually, as neuropsychologists came to believe that Freeman's basic explanations did not make scientific sense, lobotomies simply went out of style.

Freeman never gave up. Traveling around the country in his specially designed "Loboto-mobile," he performed drive-by lobotomies on new patients and checked upon

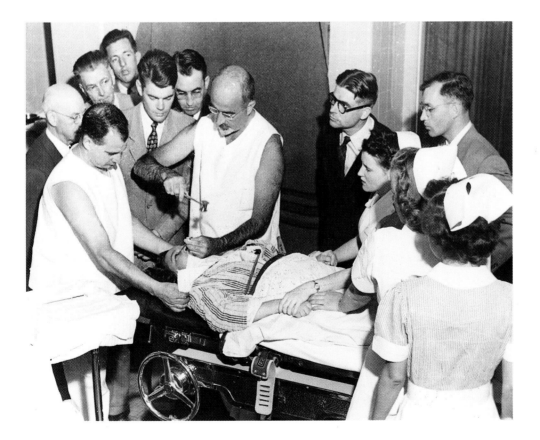

the progress of former ones. Setting aside his Parisian instruments for an icepick, he trimmed his standard operating time down to ten minutes.

"As Scientific as Physics"

All along some opponents of psychosurgery had been so disturbed by lobotomies that they edged away from Freeman and his ilk in the late 1940s and early 1950s. Psychoanalyst Dexter Bullard, who headed the private Chestnut Lodge Sanitarium in Maryland near Washington, D.C., canceled a consulting relationship with Freeman and barely spoke to him thereafter.

At small, prestigious Chestnut Lodge psychoanalysts, not administrators, were basically in charge, and the life of a patient was centered upon hour-long psychotherapy sessions given three or four times a week. Success with the "shell-shocked" victims of World War I had spurred the growth of therapy in the following decades. By the 1950s there were not only Freudian psychologists but also neo-Freudians, post–neo-Freudians, and so on without foreseeable end. Despite furious disagreements over methods of treatment, all agreed that the mind, not the brain, was the source of human behavior. It was personal experience, or nurture, not biology, or nature, that could explain our feelings, thoughts, and actions.

At Chestnut Lodge the most creative therapist was Frieda Fromm Reichmann, who typically sat down on the floor beside her schizophrenic patients and talked with them for hours. Convinced that long-term, intensive psychoanalysis was the only cure for their shadowy illness, Reichmann sought to change the "dynamic structure" of a victim's personality. Schizophrenia, she believed, was the manifestation of a severe, anxiety-provoking conflict between suppressed feelings of hostility and dependency. The conflict arose in childhood because of maternal neglect at a time when the mother's attention was indispensable to the child's survival. Reichmann thought that if she could bring out the sources of the conflict, she would be able to eliminate or modify the symptoms of mental illness.

According to critics, the work of psychoanalysts in small private hospitals was impractical as a generalized solution in huge institutions. It was also ineffective with psy-

Frieda Fromm Reichmann was convinced that intense psychotherapy conducted sympathetically over an extended period of time could cure schizophrenia, which she saw as the product of an internal conflict that arose in childhood in response to real or perceived maternal neglect. From the 1950s onward, mental health professionals began turning away from psychotherapy to psychotropic drugs such as chlorpromazine for the treatment of schizophrenics.

chotics and infamously not susceptible to scientific analysis, much less to scientific proof. Defensively many psychiatrists maintained that as one distinguished lecturer put it, "psychoanalysis is just as scientific as physics." Moreover, the psychiatric community was basically agreed that the concept of psychological determinism was ironclad: What happened in childhood absolutely determined what you became as an adult. At least one researcher published a paper showing that psoriasis, a scaly-skin disease, was caused by repression.

Whatever the conflicts at the professional level, the general public was comfortable with a homogenized understanding of the principles of psychoanalysis. It was easy to believe, even painfully obvious, that we had been produced by the sum total of our experiences. It seemed natural enough to expect that changes in behavior could be effected by talking about our thoughts and feelings.

Blaming Mom

So much of human unhappiness seemed to go back to Mother. Working with monkeys, Harry Harlow investigated just how critical the adequate mothering of infants must be

to adult development. In one experiment he found that infant monkeys, forced to find comfort with substitute inanimate mothers, preferred the dummy made of cloth to one crafted in wire. In another, he proved that baby monkeys deprived of mother love tended to develop severe behavioral problems, even autism—a complete withdrawal from reality.

Harlow's work was firmly in the 1950s mainstream, where the stress on child rearing was so prevalent in the public consciousness that many mothers felt extremely pressured to perform well. Cruelly, mothers with mentally ill children might suffer a deep sense of guilt that was often exacerbated by the treatment they received at the hands of doctors and nurses living by the day's assumptions. Some psychiatrists spoke of "refrigerator mothers." As a consequence, some anxious parents of autistic children tended to turn away from psychiatry toward education to find treatment for their troubled children. Only in the 1980s did parents in large numbers again seek help from psychiatrists.

For the moment there was no one else so readily available for blame. Because

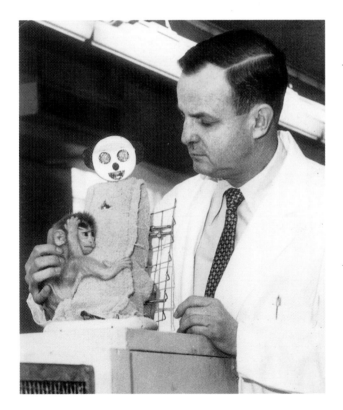

In Harry Harlow's experiments, when baby monkeys were separated from their natural mothers and offered a choice between a cloth dummy and a wire model, they eagerly clasped the softer, warmer surrogate. Monkeys entirely deprived of mothering might withdraw into an autistic state. Unfortunately, such experiments were sometimes interpreted to mean that the problems of seriously disturbed human children were inevitably caused by impersonal or frigid mothering.

neither psychotherapy nor psychosurgery was proving successful in treating psychosis, the intensifying frustration of caregivers, theorists, relatives, and patients themselves found some relief in putting the blame on moms.

"Nothing Short of a Miracle"

Then in 1952, another miracle occurred. Or was it lobotomy all over again?

"I was interested in protecting my patients from postoperative shock," neurosurgeon Henri Laborit explained later. "Before a patient is operated on, he is not very calm. Eight days before an operation, the anxiety is already building."

To relax his surgical patients at Val de Grâce Hospital in Paris, Laborit decided to try some newly developed drugs known as phenothiazines. One of them, a neuroleptic or tranquilizer called chlorpromazine (CPL), produced what he called "a profound psychic and physical relaxation." Intrigued by the strength of this result, Laborit suggested to colleagues in psychiatry that the drug might be useful with highly agitated mental patients. Indeed his intuition proved correct; the calming effect was remarkable.

These results caught the attention of U.S. psychiatrist Heinz Lehmann as he soaked in a bathtub one Sunday evening perusing medical journals. Although he was skeptical of any treatment that sounded as miraculous as CPL therapy, and also feared that the unexpected serenity could result from a kind of chemical lobotomy, he decided to find out for himself. Since these were the days before institutions typically had ethics committees to vet experiments involving human beings, he turned to his own nurses as subjects: "I gave my nurses CPL one week, barbiturates the next. Under CPL, they could perform simple tests like tapping [on cue in response to commands], but they did terribly under barbiturates."

This was cause for excitement because it suggested that the new drug could calm psychotics without inducing drowsiness. He began prescribing CPL to his disturbed patients. "After a few weeks, for many of them," Lehmann found, "there were no more hallucinations, no more delusions. They could go home. It was nothing short of a miracle."

For the first time, in other words, the symptoms of various forms of psychosis, including schizophrenia and manic depression, could be alleviated without intrusive, potentially harmful surgery. It was a phenomenal breakthrough that induced euphoria in many psychiatrists and inspired *Life* magazine to run a feature story titled "From Insanity to Sanity in 12 Days."

Syd. Edwards del.   Pub. by T. Curtis, S.t Geo: Crefcent Oct: 1.1804.   F. Sanfom sculp

But CPL did not hold the field alone. In traditional Hindu medicine a plant called *Rauwolfia serpentina* was used in the treatment of insanity. When Indian physicians studied this ancient remedy in the 1930s, they discovered that it lowered blood pressure. In turn Swiss researchers became interested in this provocative plant and isolated its active ingredient, reserpine, in order to manufacture an effective blood pressure medicine. Finally, the American psychiatrist Nathan Kline, aware that Hindu practitioners had succeeded in treating Indian psychotics, decided to try reserpine on his schizophrenic patients. Not only did they become calmer, they were also less suspicious and more cooperative.

Kline's work was widely influential, but not simply because of his demonstrated success. By coincidence, two neurotransmitters had recently been discovered in the brain: norepinephrine and serotonin. Throughout the body nerve cells release the chemical substances called neurotransmitters in order to send instructions to the next nerve cell or to a muscle or other part of the body. Without neurotransmitters we could not move or indeed perform any physical function. By the same token, neurotransmitters are essential for every iota of our thought or emotion or learning.

Reserpine, it turned out, reduced the brain's levels of norepinephrine and serotonin. In other words, the antipsychotic drug that changed behavior worked by changing brain functioning. It followed that the physical brain at last was proved to affect some aspects of human behavior.

To some observers, this discovery was regarded as possibly the greatest breakthrough in the history of the behavioral sciences. Heinz Lehmann, who was originally somewhat skeptical, has recalled the dramatic results on the institutional level:

> When I started out in psychiatry, [practitioners] thought psychosis was
> an incurable disease. Mental hospitals were like snakepits. After the drugs,
> the hospitals took on an entirely new character. . . . For the first time, it

Reserpine found naturally in snakeroot, or Rauwolfia serpentina, *is one of many mentally or physically alleviative substances provided by the plant world. It traveled an unusual arc from Hindu folk medicine to analysis in a Swiss research lab to accepted therapeutic agent in the treatment of schizophrenia. At least one quarter of prescription drugs today use chemical compounds extracted from plants.*

> *looked as if there's a possibility that severely mentally ill institutionalized patients, with appropriate treatment, might be able to be released and live in the community.*

Exhilarating as the changed outlook for treating psychotics may have been, thoughtful clinicians were unhappily puzzled. Several nettlesome questions rapidly emerged. Do all mental problems have a direct physical cause sited in the brain, and can this site always be found? The individual personality created over a lifetime can apparently be radically altered either by pill or by psychoanalysis; how is that possible?

Answers were to come only after decades of ever more spirited research and experimentation.

The Cognitive Revolution

By mid-century, many theorists who objected to the behaviorist model of mind began publishing work that led to the so-called cognitive revolution. Their thinking drew upon several earlier twentieth-century sources, including British mathematician Alan Turing, who in 1936 suggested that a calculating machine using binary code could perform an indefinite number of operations. By 1950, he had gone even further with his famous idea of the "Turing machine test": A computer could be programmed, he said, so that its answer to a question would be indistinguishable from a human's responses.

Meanwhile, Warren McCulloch and his associate Walter Pitts had been working since the 1940s on the idea that the activity of the brain's nerve cells could be understood as resembling a sequence in logical thought. Much as one idea leads to the next or does not, they reasoned, a neuron causes another neuron to fire or not. In other words, as some felt that machines could be built to imitate the brain, others were investigating just how closely the brain might resemble a powerful computer.

Also in the 1940s, Norbert Wiener had been pondering the possible similarities between the workings of the human nervous system and the connections necessary to logical thought. In 1948, he effectively announced the advent of a new science in his book *Cybernetics,* which demonstrated many possible parallels between the thinking strategies of living brains and the computations of new computers. Earlier, Claude Shannon saw that information can be understood essentially as a decision made between two equally likely alternatives. Each unit of information in this model was therefore a binary digit, or "bit." According to this information theory, which Wiener incorporated into his work, information is processed in the same way whether the transmission device is human or machine. In 1953, twenty-eight-year-old Noam Chomsky published *Syntactic Structures,* a revolutionary book that proposed that all humans are born with a "universal grammar." Language, in other words, is genetically passed down as what he called "a biological endowment."

On September 11, 1956, these and other developments important to the foundation of cognitive science were discussed at the landmark Symposium on Information Theory at MIT. For the next five years, the new movement began to bring together, as psychologist George A. Miller has written, "human experimental psychology, theoretical linguistics, and computer simulation of cognitive processes."

The human body's ten billion or so nerve cells or neurons transmit all human thought or feeling by communicating with each other over long distances through their axons, the thicker branches shown extending from neuron cell bodies here. At the terminal end, an axon may divide into as many as ten thousand separate nerve endings. A neuron's dendrites, the thinner connections visible here, receive the information projected from other neurons.

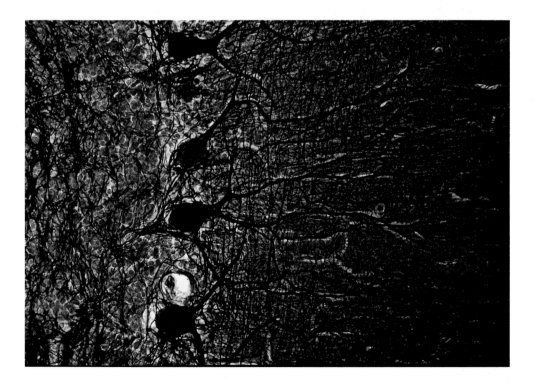

Miller and Jerome Bruner founded the Center for Cognitive Studies at Harvard in 1960, the wellspring of new developments in cognitive science for at least the following decade. No longer was the stimulus-response model of the behaviorists considered an accurate picture of the working of the brain. Instead, much like a computer, the human brain was seen as having plans or goals rather than lying passive until the outside world impinged. Certain processes of learning were seen to be universal through-out the human family, no matter what the individual's local language or cultural experience—and the patterns of these processes could be predicted. These innate patterns are interactive with the world, not reactive.

In *The Mind's New Science,* Howard Gardner set down five distinguishing features of cognitive science that were still applicable as the discipline grew and changed through the 1990s. First, cognitive scientists agree that human thought can best be understood as an independent level of representation, a preexisting "mind" that thinks in symbols or images. Second, most use computers both as a model for the working of the brain and as an important analytical tool. Third, they factor out humanistic concepts like feeling and historical context as too complex and vague to be studied scientifically at this point. Fourth, cognitive science is by nature interdisciplinary, drawing upon neuroscience and linguistics, psychology and artificial intelligence. Fifth, following the tradition of philosophy as far back as ancient Greece, Gardner believes that cognitive science is in essence concerned with the nature of knowledge.

The issues addressed by cognitive science go to the very substance of human personality and individual consciousness, but do they suggest that, after all, biology is destiny? Cognitive scientists in effect take nature's side in the continually unresolved nature/nurture controversy. The pendulum never pauses long at one end of the arc of the nature/nurture debate.

Obsessive-Compulsive Disorder

Consider a bizarre affliction that clearly brings together the influences of nature and of nurture. It is not unusual for someone to describe in casual conversation a repetitive pattern of behavior as "obsessive," as when one member of an "odd couple" accuses the other of "obsessing" about cleanliness. But the illness known as obsessive-compulsive disorder (OCD) is no situation comedy.

Sufferers of OCD are confined to a hellish mental state, even as they move about frantically performing their irresistible rituals. Why do they, for example, continually leave the room to wash their hands, which are spotlessly clean? Extremely and persistently anxious, they are fully aware at the rational level that their distress is out of all proportion to reality. Still, they may practice this unnecessary hygiene hundreds of times a day.

OCD threatens to consume its victims. Approximately five million Americans,

"Colorless green ideas sleep furiously," wrote linguist Noam Chomsky in 1957, thus gaining the attention of experts in several disciplines and illustrating a basic principle of cognitive science. Chomsky's apparently nonsensical sentence has properties that we recognize, he argued, because all humans intuitively understand the rules and principles of language, as well as other functions of the brain.

more than half of them women, spend an average of four hours a day engaged in their repetitive rituals. Throughout the world perhaps 2 percent of people from all social classes are afflicted. Until recent times the standard explanation for this behavior was demonic possession. With the advent of psychoanalytic ways of thinking late in the nineteenth century, analysts looked into the patient's childhood history for clues to the source of the affliction. But there were no dramatic instances of psychoanalytic cure.

As drugs became more common in treatment of mental disorders, antidepressants were tried with OCD patients since many of them also suffer from depression. By the late 1970s one particular antidepressant, Anafranil, seemed to be especially effective with both depression and the obsessive-compulsive behaviors. Because this drug blocks the reuptake of serotonin, thereby producing the effect of an increase in serotonin, researchers surmised that the brain's serotonin was somehow out of balance in OCD patients. In the 1980s several serotonin reuptake inhibitors became available, including the near-legendary Prozac. They often had a beneficent effect on obsessive-compulsive behaviors.

In a parallel attempt to deal with OCD, behaviorists gave patients a kind of behavioral training. If, for example, a sufferer continually washed her hands because of an unreasonable obsession with germs, she would be directed to resist her compulsion by engaging in some pleasurable or useful project for fifteen minutes. When the treatment is effective, repetition of the alternative behavior can decrease the intensity of the obsession. In the language of the behaviorist school, the OCD behaviors are thereby "extinguished."

To Scott Rauch, head of psychiatric neuroimaging research at Massachusetts General Hospital, the most seductive mystery still lay hidden somewhere within the brain's activity during these successful treatment modes. Put simply, what went on in the brain of an OCD patient as he responded either to behavioral training or to drug therapy? His study, one of several to investigate this question in the 1980s and 1990s, exemplifies an increasingly important trend in studying OCD as a sign of central nervous system dysfunction.

Studying the brain circuitry of a patient with any mental illness may not necessarily be useful if he is not exhibiting symptoms at the time. Rauch and others guessed that by evoking the symptoms, he would stimulate brain activity that would be more likely to be physiologically related to the disorder.

He worked first with a patient who had a horror of dirt. She was asked to bring two towels, one clean, one dirty, to the Mass General labs. Holding the clean towel, she was asked to inhale a radioactively tagged form of carbon dioxide that would help highlight her brain activity on a PET scan. The equipment captured an image. Twenty minutes afterward she was handed the soiled towel, which instantly triggered an overwhelming train of obsessions, including the urge to wash immediately. For a second, contrasting PET image, she inhaled the gas again.

Positron-emission tomography, or PET, is a three-dimensional imaging technique that can portray brain structures even as it measures brain activity. It works by measuring how quickly different parts of the brain consume glucose, its source of energy, and also draw upon oxygen. Seconds after a change in brain activity, specific areas will show changes in blood flow and metabolism. During different mental states different areas of the brain seem to light up. Rauch was able to conclude that the region of the brain glowing when his distressed subject touched the offensive towel was active whenever she exhibited symptoms of OCD.

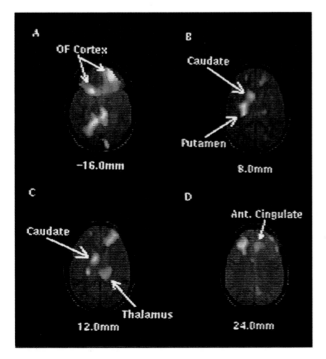

Certain areas of the brains of victims of obsessive-compulsive disorder (OCD) may be unusually stimulated when they practice their dreary rituals. In this PET (positron emission tomography) scan taken by Scott Rauch and his team in 1993, short-lived radioactive isotopes create bright spots when the brain metabolizes them. The images show the average levels of activity in the brains of a group of OCD patients when their obsessive behaviors were provoked. They suggest that OCD activity is connected with at least three specific regions of the brain.

If that assumption was accurate, OCD was linked to increased activity in two parts of the brain at once: the basal ganglia, a section associated with movement, and the limbic system, located in the prefrontal cortex. Rauch and other researchers suspected that the circuitry between the two must be impaired in OCD patients.

For comparison, Rauch studied the brain activity of normal subjects as they performed learning tasks, paying special attention to the basal ganglia because they are involved in nonconscious learning. In fact the normal subjects, as they performed repetitive movements in response to commands given on a computer screen, didn't even realize that they were learning a ritual behavior. It turned out that they and the OCD patients used different brain systems for dealing with these activities.

Perhaps more surprising, Jeffrey Schwartz and his team of psychologists and psychiatrists at the UCLA School of Medicine have discovered that behavioral therapy for OCD can affect the physical brain. Their brain-imaging shows that the four structures involved in the abnormal activity begin to behave normally in the 80 percent of cases in which behavioral therapy succeeds.

Eric Hollander, the director of the Mount Sinai School of Medicine's obsessive-compulsive disorder treatment program in New York City, has explained the significance of these results: "This tells us that effective behavioral treatments can have biological effects, not just psychological ones."

Other experimenters have also discovered connections between OCD and the activity of the brain, as well as an apparent lowering of body metabolism when the disorder is treated and a possible link between OCD and dysfunction of the brain, perhaps especially in the right hemisphere. But these potentially rewarding lines of inquiry do not seem to prove that OCD or any other mental disorder could be understood only in terms of biology. At Yale scientists specializing in genetic studies of families discovered that OCD and Tourette's syndrome, a disease manifested by uncontrollable movements and inappropriate outbursts of speech, were genetically similar.

Lab experiments revealed that similar but not the same brain systems are dysfunctional in each patient: Tourette's is mediated through that part of the brain that affects movement, while OCD is mediated through a region affecting thoughts. The genetic predisposition took two different pathways.

In sum, it appears that behavior cannot be explained by resorting only to the brain or to the mind. Both somehow work together in making us what we are.

Novelty Seeking

The line is crossed in areas where we think ourselves independent individuals, distinctive personalities. For the first time researchers recently discovered a specific gene linked to a specific personality trait—that is, to a normal attribute of character, not a mental disorder.

In the normal brain dopamine is an essential chemical communication signal. For a long time it has been implicated in impulsive, extravagant behaviors. For example, it is the dopamine system that probably responds directly to recreational drugs like cocaine, nicotine, and alcohol. People who seek out these substances are, by definition, seeking novelty. In contrast, patients afflicted with Parkinson's disease have unusually low novelty-seeking behavior, along with degeneration of the cells that produce dopamine.

To assess a physically healthy subject's actual level of novelty-seeking behavior, researchers use personality questionnaires, but they also take blood samples in order to measure a segment, or exon, of the D4 dopamine receptor gene. It turns out that there is a strong correlation between a high number of exons on the gene and quick-tempered, impetuous, fickle behavior.

Yet again it is not really that simple. When twins with identical genetic heritage are studied, tests show that about half their novelty-seeking behavior is biologically based and the other half attributable to some environmental influence not yet identified or defined.

How Many Are We?

A great deal of simplistic speculation has been written in the last years of the twentieth century about the two halves of the brain. In popular lore, the right is the seat of creativity and wonder (not to mention psychic abilities and the practice of "inner skiing"), the left the home of rationality and organization. In fact researchers have uncovered amazing but very subtle distinctions between the brain's two halves when a connecting nerve tract, the callosum, has been accidentally or surgically severed.

Elegant models of duality of function in the brain were widely published in the nineteenth century, but the concept was virtually forgotten for the first half of our century, largely because the leaders in the study of the mind believed that psychology was not neurology—i.e., personality was not derived from brain structures. The idea of a dual brain was dismissed in most quarters as "brain mythology."

Schizophrenia replaced hysteria as the most troubling of mental illnesses in the twentieth century, but even though the word itself literally means "splitting of the mind," psychiatrists did not actually picture two halves of the mind at work. Neurologists, too, believed in a single, holistic brain, and many experiments indicated that functions are not localized in a specific part of the brain. For example, brain-damaged individuals might briefly lose some vision, then regain it. Apparently, the brain was able to compensate for the loss by making use of or reorganizing undamaged areas.

Still, not everyone was convinced, and by the 1950s, many researchers were discovering that in fact the left and right hemispheres of the brain were inextricably linked to specific activities. The left was seen as "verbal" or "symbolic," the right as "visual" or "imaginative." Soon, some surprising experiments would dramatically bring to the forefront of science the likelihood that the halves of our brain are decisively, fascinatingly different.

Since the early 1940s some patients with intractable epilepsy have been treated with split-brain surgery. In studying such patients, Dartmouth College's Michael Gazzaniga discovered in the 1960s that they have separate mental systems with differing cognitive styles, though not in the popularized way. The left brain dominates in all language processes; the right is dominant in tasks involving visual construction.

For example, if a visual stimulus is presented to the left hemisphere, which is able to comprehend and speak language, a subject with a split brain can accurately describe its contents—say, a film of a villain hurling a firebomb. If the same unsettling scene is shown to the right hemisphere, the patient denies seeing it but describes emotional reactions of nervousness and alarm and misattributes this inchoate uneasiness to the experimental setting or even to the experimenter's personality. In other words, the right brain retains some of the information in some way and passes it along, though not verbally, and the left brain tries to make sense of the nonverbal information by translating it into verbal form. Gazzaniga calls this subconscious process "spurious emotional/cognitive correlation," a phenomenon that may affect those of us with whole brains in various unrecognized ways throughout our lives. One of his subjects, Gazzaniga has written, "grabbed his wife with his left hand and shook her violently, while with the right hand trying to come to his wife's aid." Can even the wisdom of Solomon sort out intent and personality here?

This apparent existence of two separate minds in one human head is a challenge to scientists and philosophers alike. Which of the two brains, say, is responsible when a decision involving morality is made?

The most important goal of Gazzaniga's continuing experiments in the 1990s is to understand the biologic nature of personal experience. When we have a specific feeling, is it sometimes based upon the right brain's inability to convey visual information in terms that the left brain can understand accurately? Is our ability to perceive reality continually filtered by the varying quality of communication between the two hemispheres of our brain? The abiding questions of who we are, how we think, what defines our personalities continue to stimulate new avenues of research, long after Charcot and Freud began their explorations into the mechanisms within. As Anne Harrington writes in *Medicine, Mind, and the Double Brain,* this is "a time when the explanatory possibilities of the brain sciences are widely perceived as almost limitless."

Answers on the Family Tree

Toward the end of the century more than one "new" disease seemed cruelly devastating, but perhaps none more than Alzheimer's. In the most common form of this disease victims begin to decay mentally after age sixty-five, slowly becoming ever more confused, irritable, forgetful, then finally unable to speak, move, or feed themselves. The rate of deterioration varies, but the patient usually dwindles through three stages, each more frustrating to him or her and each more painful to loved ones than the last.

For reasons that might often be linked to genetic inheritance, a brain protein, amyloid, changes form and accumulates in clumps in nerve cells, virtually strangling them to death. Alzheimer's cannot be reliably diagnosed until the patient dies, when an autopsy will uncover the scrambled nerves and abnormal amyloid deposits. Perhaps four million people are suffering from the disease in the United States alone, and no cure is available, though some drugs and vitamin regimens are thought to slow the progress of the disease by a few months.

One clue to the origins and behavior of this peculiar affliction has been found among poor coffee farmers in Colombia, all probably descendants from the same Spanish roots in the late eighteenth century. Not far from Medellín in the department of Antioquia, five families related by marriage fall victim to a particularly aggressive form of early-onset Alzheimer's, often showing the first symptoms when they are only in their mid-forties or even in their thirties. The average survival rate after the initial signs of deterioration is only eight years. This rare form of Alzheimer's, it turns out, is caused by a single mutated gene found on Chromosome 14. Half the children of any victim will also contract the disease because the mutated gene is dominant. In autopsies it is clear that it has stimulated the spread of a specific amyloid, A-beta-42. Sporadic or noninherited types of Alzheimer's are caused by other forms of amyloid, but lessons may be learned in South America that will explain the mechanism at work in all types of the disease. Perhaps most intriguing to researchers, a few of the doomed victims in Antioquia enjoy a natural reprieve; they do not begin to deteriorate until two decades after the average onset age of forty-seven. This unexplained delay may hold a significant clue to gaining control of the rate of deterioration or the age of first onset.

The Odyssey Continues . . .

As our century draws to a close, we know more about the diverse origins of personality than our nineteenth-century predecessors would have thought possible or would have been able to believe. Better to be demonically possessed, they might have said, than to deal unprepared with our generation's gyrating kaleidoscope of psychoanalytic theories, behaviorism, physical deformations of the brain, unbalanced body chemistry, and genetic heritage. But we are prepared, or should be, and the continuing research, the probable contradictions and ambiguities are likely to accelerate from our era into the next century. Will there be more astonishing discoveries in the days of our descendants? Will the laboratory prove, as some now suggest, that the emotional work of psychotherapy actually changes brain structure? Will the possible solutions to mental illness become even more numerous? Perhaps. Let us hope so.

This, our splendid century—beginning in despair about the healing of mental discord and ending with so many different analyses and alleviative treatments and reasons for optimism—will surely also be remembered for appreciating so well at last, and for digging so deeply, into the regions of consciousness where brain and mind in subtle, abiding ways perpetually overlie. The human brain hungers. We are what we know.

According to the recent theory of "neural Darwinism," as developed by Gerald Edelman, the human brain does not appear out of the womb as an exactly preordained, hard-wired device. It takes shape as the child interacts with a caregiver. Loving communication between mother and child helps shape our ability to understand the world by reinforcing neuronal connections in the brain that promote effective thinking and perception while allowing inessential connections to wither from disuse.

ILLUSTRATION CREDITS

P. ii: Robert William and the Hubble Deep Field Team (STScl) and NASA; p. viii: Felice Frankel and Philip Sharpe; p. x: AP/Wide World Photo.

Chapter One

P. xviii: Tony and Daphne Hallas, Astro Photo; p. 3: Lowell Observatory; p. 5: The Observatories of the Carnegie Institution of Washington; p. 6: Brown Brothers; p. 8: The Observatories of the Carnegie Institution of Washington; p. 10: Brown Brothers; p. 13: The Observatories of the Carnegie Institution of Washington; p. 14: Harvard College Observatory/Science Photo Library; p. 15: The Observatories of the Carnegie Institution of Washington; p. 16: Harvard College Observatory; p. 18: Albert Einstein, licensed by the Hebrew University, represented by The Roger Richman Agency, Beverly Hills, CA; p. 23 left: Culver Pictures, Inc.; p. 23 right: Michael Gilbert/Science Photo Library; p. 26: AIP Emilio Segrè Visual Archives, Margrethe Bohr Collection; p. 27: The Niels Bohr Archive, Copenhagen; p. 29: photo by Francis Simon, courtesy AIP Emilio Segrè Visual Archives; p. 30: Corbis-Bettmann; p. 32: photo by Paul Ehrenfest, courtesy AIP Emilio Segrè Visual Archives; p. 35: Ernest Orlando Lawrence Berkeley National Laboratory, University of California; p. 36: University of California Lawrence Radiation Lab, courtesy

AIP Emilio Segrè Visual Archives; p. 39: Los Alamos National Laboratory; p. 42: Francois Gohier/Photo Researchers, Inc.; p. 42 inset: courtesy NRAO/AUI; p. 44: Jonathan Blair/National Geographic Image Collection; p. 45: John Hutchings (Dominion Astrophysical Observatory), Bruce Woodgate (GSFC/NASA), Mary Beth Kaiser (Johns Hopkins University), and the STIS Team; p. 48: property of AT&T Archives; p. 50: Michael Gilbert/Science Photo Library; p. 52: Science Photo Library/Photo Researchers, Inc.; p. 54: CERN Photo; p. 55: CERN Photo; p. 58: Dr. Seth Shostak/Science Photo Library/Photo Researchers, Inc.; p. 60: Explorer/Photo Researchers, Inc.

Chapter Two

P. 62: The Huntington Library, San Marino, California; p. 64: Culver Pictures, Inc.; p. 66: National Air and Space Museum, Smithsonian Institution, Photo No. A-38681; p. 68: Brown Brothers; p. 70: National Air and Space Museum, Smithsonian Institution, Photo No. A-2466A; p. 71: Image Select International London; p. 73: Brown Brothers; p. 74: Corbis-Bettmann; p. 78: National Air and Space Museum, Smithsonian Institution, Photo No. 97-16136; p. 80: Brown Brothers; p. 82: Corbis-Bettmann; p. 84: Corbis-Bettmann; p. 85: Image Select International London; p. 86: Corbis-

Bettmann; p. 87: Brown Brothers; p. 89: UPI/Corbis-Bettmann; p. 91: Archive Photos; p. 94: unknown; p. 97: Collection of The New-York Historical Society; p. 99: Gottscho-Schleisner Museum of the City of New York/Archive Photos; p. 101: UPI/Corbis-Bettmann; p. 105 top: Science Photo Library/Photo Researchers, Inc.; p. 105 bottom: Archive Photos; p. 106: Hagley Museum and Library; p. 110: IBM Corporation; p. 111: courtesy of MITRE Archives; p. 113: IBM Corporation; p. 115: Archives, U.S. Space and Rocket Center; p. 116: AT&T Archives; p. 119: sovfoto/eastfoto; p. 120: NASA; p. 122: NASA.

Chapter Three

P. 126: G. Brad Lewis/Photo Resource Hawaii; p. 128: Rod Planck/Photo Researchers; p. 130: Corbis-Bettmann; p. 131: D. J. Roddy and K. Zeller, U.S. Geological Survey; p. 133: The Hulton Getty Picture Collection/Tony Stone Images; p. 134: AIP Emilio Segrè Visual Archives, William G. Myers Collection; p. 137: Ann Ronan at Image Select; p. 139: The National History Museum, London; p. 141: Ann Ronan at Image Select; p. 142: Alfred Wegener Institute for Polar and Marine Research; p. 145: Princeton University Library; p. 148: Bruce Heezen and Marie Tharp; p. 148 inset: Lamont Geological Observatory; p. 150: David Hardy/Science Photo Library; p. 151: Kevin Shafer/Peter Arnold, Inc.; p. 154: Brown Brothers; p. 158: Peter Kain/Sherma B.V.; p. 161: K. Cannon-Bonventre/Anthro-Photo; p. 161 inset: Bob Campbell; p. 162: John Reader/Science Photo Library; p. 163: Institute of Human Origins/Dr. Donald Johanson; p. 167: David Scharf/Peter Arnold, Inc.; p. 168: by permission of King's College, London; p. 169: Archive Photos; p. 172: Fred Bavendan/Peter Arnold, Inc.;

p. 174: Norbert Wu/Peter Arnold, Inc.; p. 176: BioPhoto/Photo Researchers, Inc.

Chapter Four

P. 178: courtesy of the Archives of the Buffalo General Hospital; p. 181: Underwood & Underwood/Corbis-Bettmann; p. 183: courtesy of Library of Congress; p. 184: National Archives (90-G-124-479); p. 185: CNRI/Science Photo Library; p. 187: Culver Pictures, Inc.; p. 188: City of Philadelphia, Department of Records, City Archives, RG 78, Print 9531; p. 190: *Die Vitamine: ihre Bedeutung für die Physiologie und Pathologie mit besonderer Berücksichtigung* by Casimir Funk; p. 192: courtesy of Library of Congress; p. 194: Parke-Davis Division of Warner-Lambert, Morris Plains, New Jersey; p. 195: Sharp family; p.197: Banting Papers, Thomas Fisher Rare Book Library, University of Toronto; p. 200 left: UPI/Corbis-Bettmann; p. 200 right: Corbis-Bettmann; p. 203: Thomas Fisher Rare Book Library, University of Toronto; p. 204: Thomas Fisher Rare Book Library, University of Toronto; p. 209: Audio Visual Services, Imperial College School of Medicine at St. Mary's, London; p. 210: Corbis-Bettmann; p. 212: Sir William Dunn School of Pathology; p. 216: courtesy of the Medical Archives, New York Hospital–Cornell Medical Center; p. 218: The Hulton Getty Picture Collection/Tony Stone Images; p. 220: Biozentrum/Science Photo Library; p. 222: UPI/Corbis-Bettmann; p. 225: Manuscripts Division, University of Utah Libraries; p. 227: courtesy of Brigham and Women's Hospital; p. 230: Express News/Archive Photos; p. 231: NIH/Science Source; p. 233: Alfred Pasieka/Science Photo Library; p. 236: Baxter Healthcare Corporation; p. 239: NIBSC/Science Photo Library; p. 241: John Moss/Photo Researchers.

Chapter Five

P. 244 top left: *Nouvelle Iconographie de la Salpêtrière* 890/Boston Medical Library in the Francis Countway Library of Medicine; p. 244 top right: *Iconographie photographique de la Salpêtrière*/Boston Medical Library in the Francis Countway Library of Medicine; p. 244 bottom left: Jean-Loup Charmet; p. 244 bottom right: *Iconographie photographique de la Salpêtrière,* Bibliothèque Charcot, la Salpêtrière; p. 246: NMPFT/Science & Society Picture Library; p. 248: *A Practical Treatise on the Medical and Surgical Uses of Electricity,* by G. M. Beard; Science and Society Picture Library; p. 249: The Bakken Library and Museum; p. 251: Jean-Loup Charmet; p. 252: *Iconographie photographique de la Salpêtrière,* Bibliothèque Charcot, la Salpêtrière; p. 253: AP-HP; p. 255: A. W. Freud et al. by arrangement with Mark Paterson and Associates; p. 257: Archives of the Sigmund Freud Museum; p. 259: Edmund Engelman; p. 262: ET Archive, London; p. 264: *Memoirs of the National Academy of Sciences,* Volume XV; p. 266: Corbis-Bettmann; p. 268: Archives of the History of American Psychology at the University of Akron; p. 270: American Philosophical Society; p. 272: Professor Ben Harris; p. 276: Charles Lord; p. 279: UPI/Corbis-Bettmann; p. 281: Joanne Greenberg; p. 282: Harlow Primate Laboratory/University of Wisconsin; p. 284: The LuEsther T. Mertz Library of The New York Botanical Garden, Bronx, New York; p. 287: Manfred Kage/Peter Arnold, Inc.; p. 288: courtesy of MIT Department of Linguistics and Philosophy; p. 291: S. L. Rauch et al.; p. 297: Laura Dwight/Corbis.

INDEX

Page numbers in *italics* refer to illustrations.